Advanced Stability

published by

THE UNIVERSITY OF TEXAS

CONTINUING EDUCATION

PETROLEUM EXTENSION SERVICE

2006

Library of Congress Cataloging-in-Publication Data

Advanced stability—1st ed.
p. cm.
ISBN 0-88698-215-4
1. Offshore structures—Design and construction. 2. Structural stability.
3. Drilling platforms
TC1665.A37 2005
627'.98—dc 22 2005014417

First Edition 2006
Printed in the United States of America

Catalog No. 1.12010
ISBN 0-88698-215-4

TABLE OF CONTENTS

OBJECTIVES AND CONTENTS

1.1 DESCRIPTION

STCW95—Resolution A.891(21) specifies the knowledge, understanding, and proficiency for stability on offshore drilling units required for the function of offshore installation manager (OIM), barge supervisor (BS), and ballast control operator (BCO) on board mobile offshore units (MOUs).

The aim is to cover the entire theory of stability up to and above the standard as required by the Resolution A.891(21).

It is advisable for any person *without* a marine education or a bachelor degree in science first to read *Comprehensive Stability*, available from Petroleum Extension Service, 512-471-5940 or 1-800-687-4132.

A person with an unlimited marine deck license (master, mate) or with a bachelor degree in science should be able to read and comprehend *Advanced Stability* without first reading *Comprehensive Stability.*

1.2 CONTENT

Advanced Stability covers the fundamental stability theory in eight chapters with individual exercises for each chapter. The instruction, evaluation, and exercises cover the following subjects:

- Units of measurements, definitions, and basic stability
- Environmental forces
- Hydrostatic properties and daily stability calculations
- The mooring system
- Stability at large angles—stability curves
- Damage stability
- The ballast system
- Design and construction

1.3 OBJECTIVES

The reader should be able to:

- Calculate with formulas as set out in the section with definitions on how to use the Imperial and metric systems for forces, moments, and weight changes.
- Understand and explain the various categories of stability.
- Understand and explain the interrelation between the points G, B, and M.
- Understand and calculate the change of draft, trim, and heel caused by weight changes.
- Understand and calculate the position of G, B, and M for various types of drilling units in vertical, transverse, and horizontal configuration.
- Be able to use the Hydrostatic Property tables for multiple weight changes to find the positions of G, new draft readings, and corresponding maximum allowed VCG value.
- Understand the importance of free surface effect and perform calculations to show the reduced effect on the stability.

- Understand the purpose of the inclining test. Demonstrate with exercises how the inclining test is performed and calculated.
- Demonstrate with examples and calculation the information obtained from the stability curves.
- Understand and explain the difference between stability for small and large angles.
- Understand the basics of damage stability and the primary counteraction to be followed.
- Understand and explain the international regulations concerning stability for normal operations, survival, and damaged stability conditions.
- Understand and explain the difference of static and dynamic stability.
- Understand and use of the maximum allowable VCG curve in conjunction with the GM value.
- As an overall result, the student should be able to demonstrate with exercises how to perform a stability calculation with multiple load changes in the vertical, transverse, and horizontal configuration with the use of all the information explained in the course.
- Understand and explain the daily stability calculations including the effect of the anchor mooring system.
- Be able to calculate the various sections and forces within the catenary mooring system.
- Understand and explain the causes and consequences of damage stability.
- Demonstrate with calculations the effect and countermeasures for damage stability conditions.
- Understand and explain the principles of the ballast system.
- Understand and explain the basic principles of design and structure of the offshore drilling rigs.

INTRODUCTION

The International Maritime Organization (IMO) adopted the International Convention on Standards of Training, Certification and Watchkeeping for Seafarers (STCW) in 1978, with Amendments in 1995 and 1997, to set qualification standards for masters, officers, and watch personnel on seagoing merchant marine vessels.

Because the STCW was developed for seagoing merchant marine vessels, the IMO adopted on 25 November 1999 Resolution A.891(21) to cover the training of personnel on mobile offshore units (MOUs). Resolution A.891(21) specifies the minimum standards of competence for the functions of the offshore installation manager (OIM), the barge supervisor (BS), ballast control operator (BCO), and the maintenance supervisor. These minimum standards of competence include knowledge, understanding, and proficiency of stability.

Table 1.0 is a summary of the stability knowledge, understanding, and proficiency required in accordance with the Resolution A.891(21) for the offshore installation manager, the barge supervisor, and the ballast control operator on mobile offshore units versus the contents of the *Advanced Stability*.

TABLE 1.0—STCW REQUIREMENTS

Table with Indication of Training in Accordance with A.891(21) Covered in *Comprehensive Stability* and *Advanced Stability* for OIM, BS, BCO

<table>
<tr><th colspan="2">Knowledge Understanding and Proficiency as Required by A.891(21)</th><th>OIM</th><th>BS</th><th>BCO</th><th>Covered in Comprehensive Stability Chapter–Section</th><th>Covered in Advanced Stability Chapter–Section</th></tr>
<tr><td colspan="2">Knowledge of and ability to apply relevant international and national standards.</td><td>×</td><td>×</td><td>×</td><td>Chapt. 1, 1.1–1.3,
Chapt. 8, 8.1–8.11</td><td>Chapt. 1, 1.2,
Chapt. 2, 2.1–2.18,
Chapt. 5, 5.1–5.20,
Chapt. 6, 6.1–6.13</td></tr>
<tr><td colspan="2">Use of loading stability information from stability and trim diagrams, MOM, and/or computerised loading/stability programs.</td><td>×</td><td>×</td><td>×</td><td>Chapt. 1, 1.3,
Chapt. 5, 5.1–5.12,
Chapt. 6, 6.1–6.6,
Chapt. 9, 9.1–9.5</td><td>Chapt. 1, 1.39–1.57,
Chapt. 3, 3.1–3.19</td></tr>
<tr><td colspan="2">Understanding of fundamental principles-theories-factors affecting trim and stability to preserve trim and stability and measures to preserve trim and stability.</td><td>×</td><td>×</td><td></td><td>Chapt. 1, 1.1–1.3,
Chapt. 2, 2.1–2.16,
Chapt. 3, 3.1–3.4,
Chapt. 5, 5.1–5.12,
Chapt. 6, 6.1–6.6</td><td>Chapt. 1, 1.1–1.57,
Chapt. 3, 3.1–3.18,
Chapt. 6, 6.12–6.13</td></tr>
<tr><td colspan="2">Static and dynamic stability criteria for MOUs, environmental limits, and criteria for survival conditions.</td><td>×</td><td>×</td><td>×</td><td>Chapt. 4, 4.1–4.8,
Chapt. 8, 8.1–8.11</td><td>Chapt. 2, 2.1–2.18,
Chapt. 5, 5.1–5.20</td></tr>
<tr><td colspan="2">Understanding of inclining experiment, deadweight, and their use.</td><td>×</td><td>×</td><td>×</td><td>Chapt. 1, 1.3,
Chapt. 7, 7.1–7.5</td><td>Chapt. 1, 1.59</td></tr>
<tr><td colspan="2">Use of daily loading calculations.</td><td>×</td><td>×</td><td>×</td><td>Chapt. 5, 5.1–5.12,
Chapt. 8, 8.1–8.11</td><td>Chapt. 3, 3.1–3.19</td></tr>
<tr><td rowspan="5">Knowledge of the effect</td><td>1) Trim and stability of MOUs in event of damage and consequent flooding and countermeasures.</td><td>×</td><td></td><td></td><td>Chapt. 8, 8.2, 8.9–8.11</td><td>Chapt. 6, 6.1–6.13</td></tr>
<tr><td>2) Of loading supplies and ballasting in order to keep the unit's stress within the acceptable limits.</td><td>×</td><td></td><td></td><td>Chapter 9, 9.4
Chapter 11, 11.1–11.15</td><td>Chapt. 3, 3.14–3.19
Chapt. 8, 8.1–8.15</td></tr>
<tr><td>3) Mooring system and mooring line failures.</td><td>×</td><td></td><td></td><td>Chapt. 9, anchor system correction only in 9.3</td><td>Chapt. 4, 4.1–4.18</td></tr>
<tr><td>4) Preloading and leg stresses on JUs</td><td>×</td><td></td><td></td><td>Chapter 11, 11.1–11.15</td><td>Chapt. 8, 8.1–8.15</td></tr>
<tr><td>5) Loss of buoyancy</td><td>×</td><td></td><td></td><td>Chapter 10, 10.1–10.13</td><td>Chapt. 6, 6.1–6.13</td></tr>
<tr><td rowspan="3">Knowledge of</td><td>1) The emergency response for flooding due to damage, fire fighting, loss of buoyancy, and the effect on trim and stability.</td><td>×</td><td>×</td><td>×</td><td>Basics only in Chapt. 8, 8.9–8.10</td><td>Chapt. 6, 6.1–6.13</td></tr>
<tr><td>2) Countermeasures for damage stability.</td><td>×</td><td>×</td><td>×</td><td>Basics only in Chapt. 8, 8.9–8.10</td><td>Chapt. 6, 6.12–6.13</td></tr>
<tr><td>3) Effectively communicate stability-related information.</td><td>×</td><td>×</td><td>×</td><td>Chapt. 9, 9.1–9.5</td><td>Chapt. 1, 1.4–1.64,
Chapt. 3, 3.1–3.19,
Chapt. 4, 4.18,
Chapt. 5, 5.1–5.20,
Chapt. 6, 6.12–6.13</td></tr>
<tr><td colspan="2">The effect of trim and stability of cargo and cargo operations.</td><td>×</td><td>×</td><td></td><td>Chapt. 1, 1.3,
Chapt. 3, 3.1–3.4,
Chapt. 5, 5.8–5.12,
Chapt. 9, 9.1–9.5</td><td>Chapt. 1, 1.23,
1.38–1.42, 1.47–1.63,
Chapt. 3, 3.13–3.19,
Chapt. 5, 5.19–5.20</td></tr>
</table>

CHAPTER 1
Units of Measurement, Definitions, and Basic Stability

1.1 INTRODUCTION

This chapter is a summary to refresh the knowledge of the units of measurement, definitions, and basic stability as discussed in *Comprehensive Stability*.

1.2 UNITS OF MEASUREMENT IN THE METRIC AND CONVENTIONAL SYSTEMS

A unit is a standard measure of quantity such as length, mass, energy, etc. Compared to the SI (Système International d'Unités) or metric system, the conventional system (also called Imperial or U.S. system) is more complicated because several units are used for each quantity. For example, measurements of length can be expressed in miles, yards, feet, and inches. In the SI or metric system, only one unit is used for each basic quantity; i.e., the metre is the basic unit of length and the kilogram is the basic unit of weight. For stability calculations, both systems are used in the offshore drilling industry.

The basic units and derived quantities for each system are shown in Table 1.1.

TABLE 1.1

Fundamental Quantities and Derived Quantities	Conventional System		Metric (SI) System	
	Unit	Symbol	Unit	Symbol
Acceleration	feet per sec per sec	ft/sec^2	metre per sec per sec	m/sec^2
Area	square feet	ft^2	square metre	m^2
Density	pound per cubic ft	lb/ft^3	kilogram per cubic m	kg/m^3
Force	pound force	lb_f	Newton	N
Frequency	hertz	Hz	hertz	Hz
Length	foot	ft	metre	m
Mass	pound	lb	kilogram	kg
Power	foot-pound per sec horsepower watt	ft-lb/sec hp W	joule per second kilogram-metre per sec watt	J/s kg.m/sec W
Pressure	pound per square inch	psi	pascal (N/m^2) kilogram/square cm	Pa kg/cm^2
Temperature	degree Fahrenheit	°F	degree Celsius	°C
Tons	short tons (2,000 lb)	st	tonne (1,000 kg)	t
Velocity	foot per second	ft/sec	metre per second	m/s
Volume	cubic foot gallon barrel	cu.ft or ft^3 gal bbl	cubic metre	m^3
Work	foot-pound	ft-lb	joule Newton-metre	J N-m

BASIC UNITS

1.3 QUANTITIY AND MEASUREMENTS FOR STABILITY CALCULATIONS

LENGTH

The basic unit of length is the metre or foot.
1 metre (m) = 10 decimetres (dm) = 100 centimetres (cm) = 1,000 millimetres
1 foot (ft) = 12 inches (in.)
1 metre = 3.281 ft = 39.37 in.
1 yard = 3 ft = 0.9144 m
1 ft = 0.305 m

AREA

For metric units in most cases the area is measured in m^2.
For conventional units the area is measured in ft^2 or $in.^2$.
1 m^2 = 10,000 cm^2 = 10.76 ft^2
1 $yard^2$ = 9 ft^2 = 0.836 m^2
1 ft^2 = 144 $in.^2$ = 0.093 m^2

VOLUME

Volume is measured in m^3 or in ft^3.
1 m^3 =10^6 cm^3
1 m^3 = 35.31 ft^3 = 264.12 gal = 6.29 bbl
1 litre = 1,000 cm^3 = 0.264 gal
1 ft^3 = 1,728 $in.^3$ = 0.028 m^3
1 gal (U.S.) = 0.0038 m^3 = 0.134 ft^3
1 barrel (bbl) = 0.159 m^3 = 5.61 ft^3 = 42 gal

WEIGHT

The basic units for weight are the kilogram (kg) and pound (lb).
1 kg = 1,000 grams
1 kg = 2.2046 lb
1 litre of fresh water weighs 1 kg
1 ft^3 of fresh water = 62.5 lb versus 1 ft^3 of seawater = 64 lb
1 lb = 16 ounces (oz) = 0.454 kg
1 metric tonne (t) = 1,000 kg
1 m^3 fresh water = the weight of 1,000 kg
1 metric tonne = 2,204.6 lb = 1.1023 st
1 short ton (st) = 2,000 lb = 0.9072 t
1 long ton = 2,240 lb = 1,016 kg

FORCE AND MASS

1 Newton (N) = 1 kg.m/sec^2 = 0.2248 pound-force (lb_f)
1 pound-force (lb_f) = lb_m *32.15 ft/s^2 = 4.448 Newton
1 kg = 2.2046 pound-mass (lb_m). 1 lb_m = 0.454 kg

1.4 GRAVITY, WEIGHT, AND MASS

Gravity is the force that tends to draw all bodies towards the center of the earth.

Mass is the physical quantity of matter in a body.

Weight is the vertical force experienced by a mass as a result of the gravity force. Weight can be considered to be proportional to mass at the earth's surface.

1.5 FORCE

Force is the push or pull on a body. It is the cause of the motion of a body. By definition, force is the product of mass and acceleration.

In the metric system, the unit of force is the *newton* (*N*). One newton is the force required to accelerate a mass of 1 kilogram at a metre per second squared (kgm/s^2).

The unit of force for the conventional system is the *pound force*; that is, the force required to accelerate a mass of 1 pound at 32.15 feet per second squared (32.15 ft/s^2).

1.6 VECTOR

Vector is a variable quantity, such as force, that has magnitude and direction. A vector may be graphically represented by a straight-line arrow with a direction and a length to indicate the magnitude (fig. 1.1).

1.7 RESULTANT FORCE

The combined effect of two or more forces acting on a body is called the *resultant force*. To know the details of forces, we need to know the point of application, the direction, and the magnitude. This is graphically displayed in Figure 1.1.

The various scenarios of two forces acting on a body are shown in Figure 1.2. The resultant force of two forces acting in the same direction is the sum of the two forces. The resultant force of two forces acting opposite each other in a straight line is the difference between the two forces.

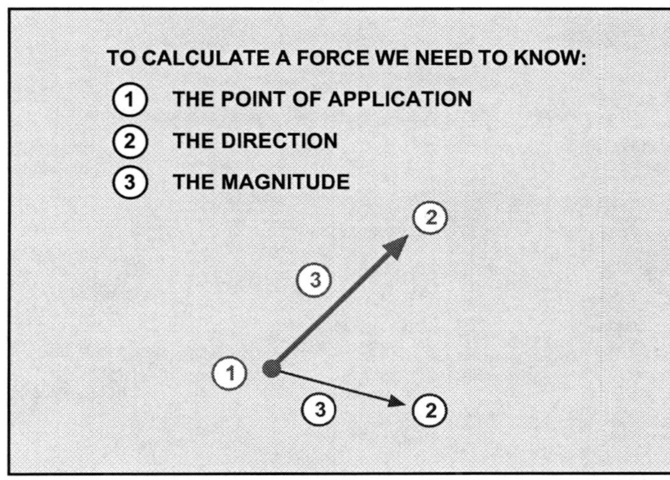

Figure 1.1 Forces

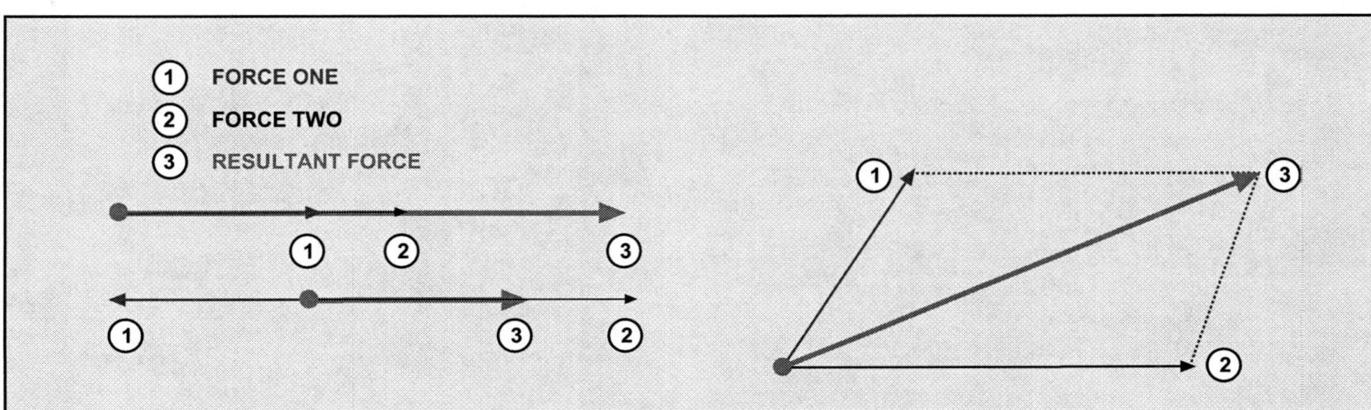

Figure 1.2 Resultant force

Examples of Resultant Force

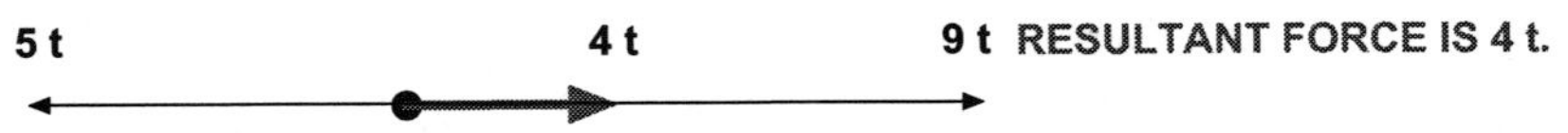

1.8 MOMENT OF FORCES

A force *F* applied to a body at a distance *d* from its axis or point *A* results in a rotation about the axis.

1.9 MOMENT

Moment is the tendency to produce rotation by a force *F* at a distance *d* about a point or axis. The effect of the rotation depends on:

1. The magnitude of the force (*F*).
2. The length *d* of the lever or arm (fig. 1.3).

MOMENT = FORCE × DISTANCE
M = F × d. SUBSTITUTE F FOR WEIGHT (W)
M = W × d

Moments can be applied about any imaginary point. This principle is important and is used as the basics for stability calculations. Stability calculations do not use just one moment but a combination of many moments.

1. The resultant moment is the total effect of all the combined moments.
2. The rotation direction about the point or axis is clockwise or anticlockwise.
3. To find the resultant, all clockwise moments are added together and deducted from the sum of all the anticlockwise moments.
4. To calculate the moment of force, any point can be used as a reference point to find the resultant moment.

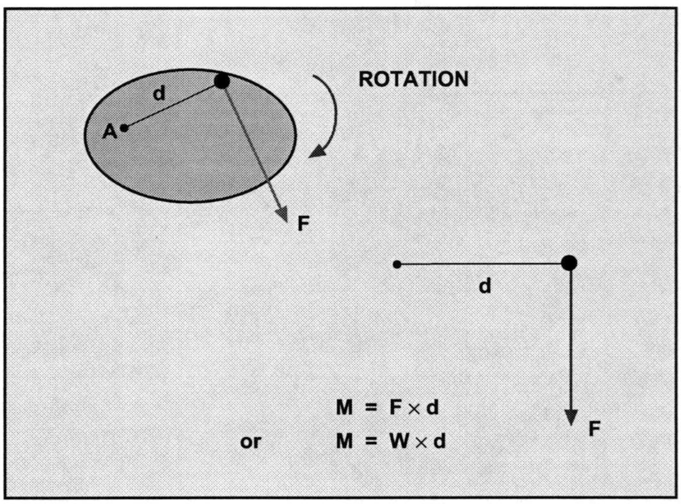

Figure 1.3 Moment of forces

1.10 THE CENTER OF GRAVITY (G)

1. The *center of gravity* (*G*) is the geometrical center of a homogeneous body.
2. The force of gravity acts through the center of gravity.
3. The force of gravity acts vertically downward.
4. The force of the gravity is equal to the weight of the body.
5. The center of gravity is the point about which a body will balance.

The position of *G* for various areas and shapes are shown in Figure 1.4.

The position of *G* can be expressed in the vertical, transverse, and longitudinal plane with respect to any reference point or line.

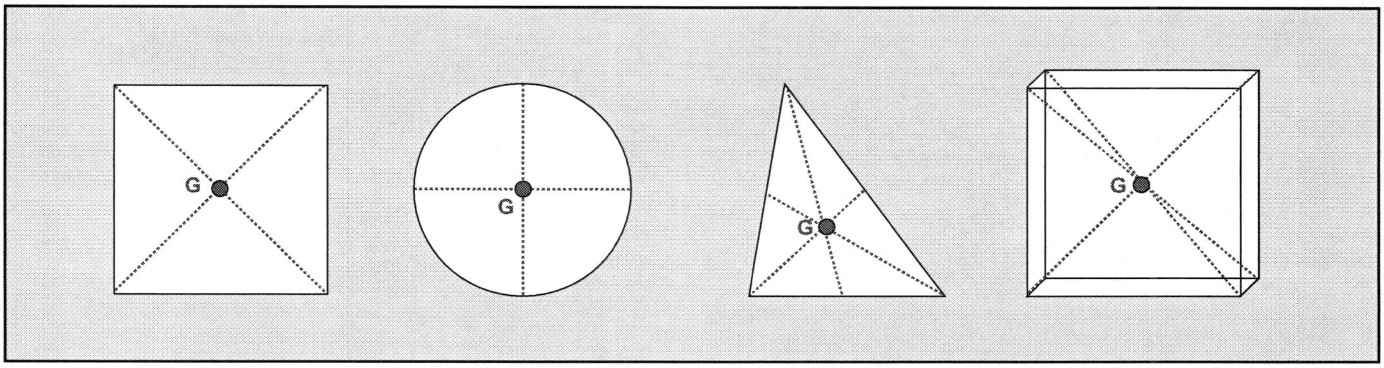

Figure 1.4 Center of gravity of homogeneous body

The *center of gravity* (*G*) or *vertical center of gravity* (*VCG*) of a rig or vessel is the point through which the *total weight of the rig* (*W*) is considered to act in a vertically downwards direction (fig. 1.5). Because the rig is not a homogeneous body, the position of the center of gravity depends on the shape and steel structure of the hull and the position of the weights stored on the rig.

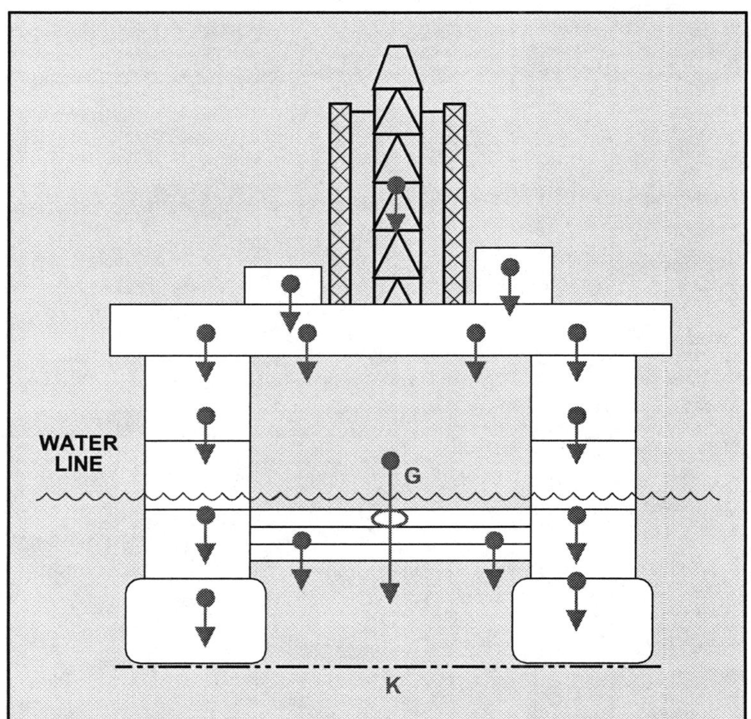

Figure 1.5 The position of G

KG is the vertical distance between the center of gravity of the rig and the keel. The keel is the base line. To know the exact position of *G* in a three-dimensional configuration, the location of *G* is indicated as an example in Figure 1.6:

1. *KG* or *VCG* is the vertical distance of *G* above the keel
2. *TCG* is the transverse distance of *G* from the longitudinal centerline of the rig. *TCG* is positive (+) to starboard and negative (–) to port.
3. *LCG* is the longitudinal distance of *G* from the transverse centerline of the rig. *LCG* is positive (+) forward of the transverse centerline and negative (–) aft.

The positive and negative configurations can be different from one rig to another.

In Figure 1.6 *TCG* is negative (-) and *LCG* is positive (+).

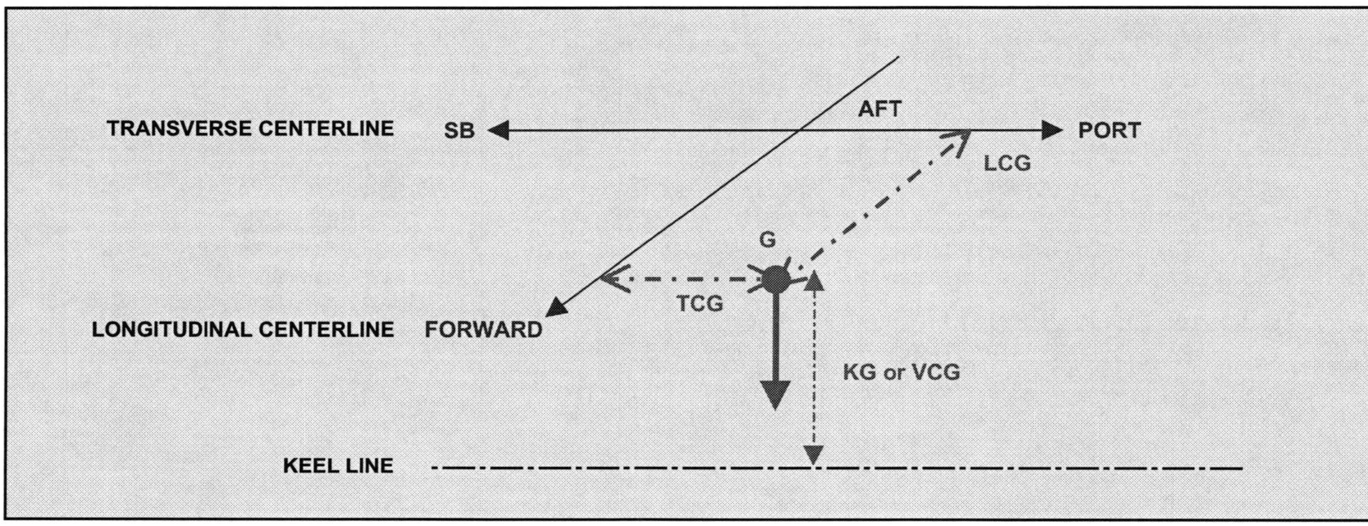

Figure 1.6 Orientation of G

Example 1. Find the distance of the position of the fulcrum from the left side of the body.

The same principle of the seesaw is used. Because *G* is the point about which the body will balance, *G* can be considered to be the position of the fulcrum.

Find the distance (*d*) of the center of gravity from the left side for a body with various weights.

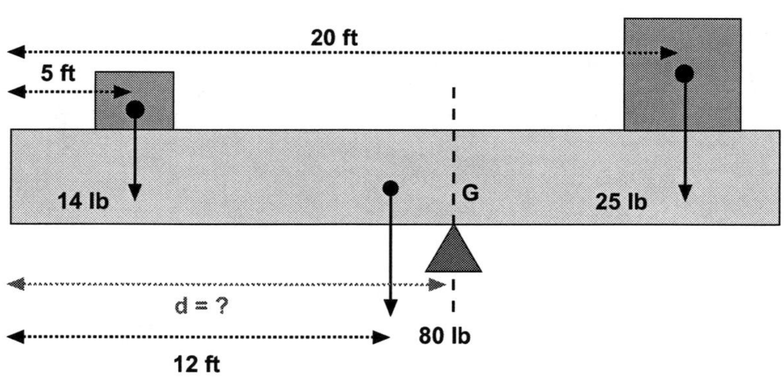

$$\text{Total moments clockwise: } M_{cw} = (14 \text{ lb} \times 5 \text{ ft}) + (80 \text{ lb} \times 12 \text{ ft}) + (25 \text{ lb} \times 20 \text{ ft}) = 1{,}530 \text{ lb·ft}$$

$$\text{Total moments anticlockwise: } M_{aw} = (14 \text{ lb} + 80 \text{ lb} + 25 \text{ lb})\, d = 119\, d$$

Clockwise and anticlockwise moments are equal.

$$119\, d = 1{,}530 \text{ lb·ft}$$

$$d = 12.86 \text{ ft}$$

Example 2. Find the position of G on a vessel with various weights.

K is the keel of the vessel. *KG* is the vertical distance from keel to *G*. *KG* is often referred to as the vertical center of gravity (VCG).

A square-shaped barge has a weight of 3,500 t lightship displacement and has a corresponding lightship *KG* of 5.5 m.

A load of 120 t is placed at 2.3 m above the keel and a second load of 400 t is loaded 6.1 m above the keel. Find the new *KG*.

The reference point to find the vertical distance is the keel.

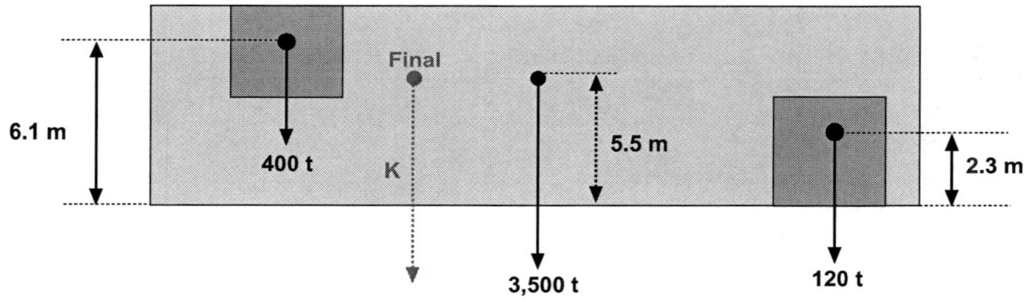

$$KG = \frac{\text{Total moment of all weights in tm}}{\text{The total moment of all weights in t}}$$

$$KG = \frac{(3{,}500 \text{ t} \times 5.5 \text{ m}) + (400 \text{ t} \times 6.1 \text{ m}) + (120 \text{ t} \times 2.3 \text{ m})}{3{,}500 \text{ t} + 400 \text{ t} + 120 \text{ t}}$$

$$KG = \frac{21{,}966 \text{ tm}}{4{,}020 \text{ t}}$$

$$KG = 5.46 \text{ m}$$

EXERCISE 1
Chapter 1

1. Define the center of gravity.

2. Explain the difference between mass and weight.

3. Calculate the resultant force angle and value for the following examples.

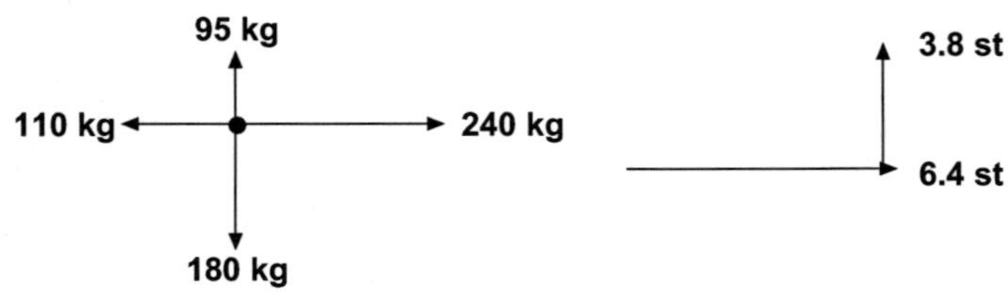

Find the unknown weight or distance for equilibrium.

4.

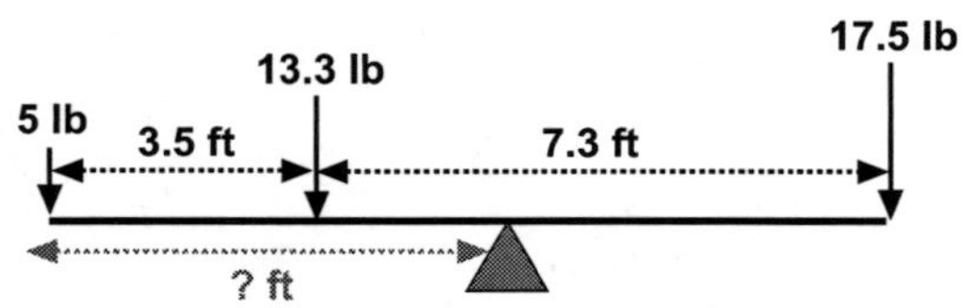

5.

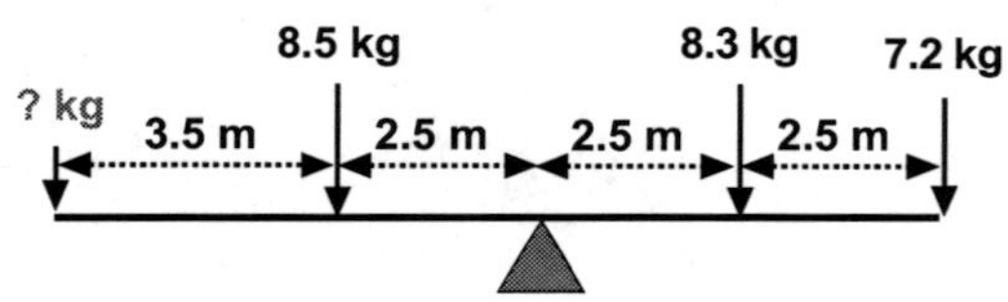

6.

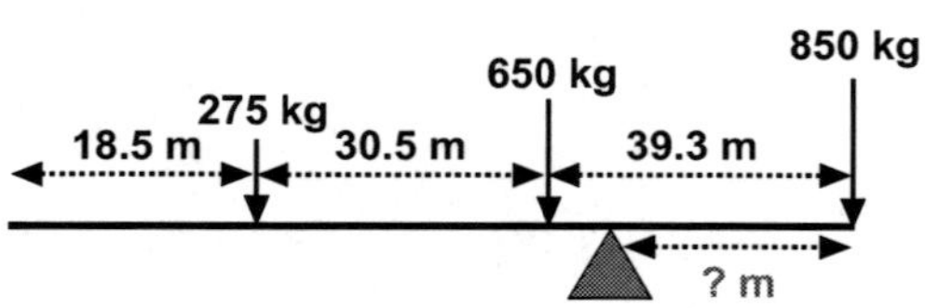

Find the resultant moment about the fulcrum for equilibrium.

7. Find the distance *LCG* for equilibrium.

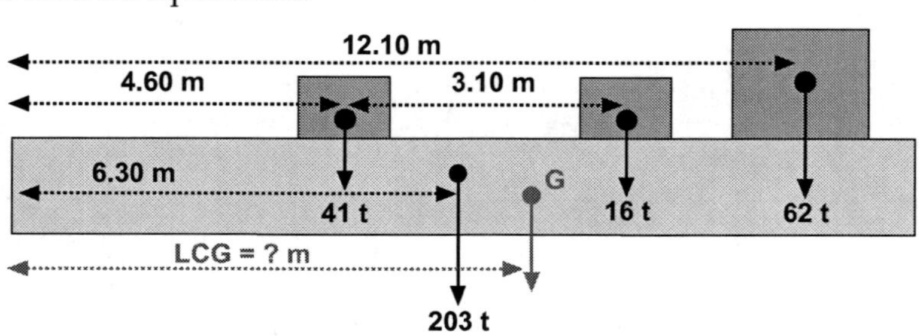

1.11 THE CENTER OF BUOYANCY

1. The *center of buoyancy* (*B*) is the geometrical center of the underwater shape of the rig (fig. 1.7).
2. The buoyancy force is the resultant of the vertical upward force equal to the weight of the displaced water. This tendency to keep the rig afloat is called buoyancy.
3. The buoyancy force acts through *B*.

Because the underwater shape is known before the rig is built, it is possible for the naval architect to calculate the position of *B* in relation with *K* for a range of draft. The results are shown in the hydrostatic properties of the rig.

Like the position of G, the position of *B* is also expressed in the vertical plane VCB, longitudinal plane LCB, and transverse plane TCB. The example shows the position of VCB or *KB*.

In the upright position (zero degree inclination), the location of *B* will be on the centerline of the rig. Any heel or trim moves *B* to one side and away from the centerline because the shape of the underwater volume changes. The buoyancy force through *B* remains vertical and upward (fig 1.8).

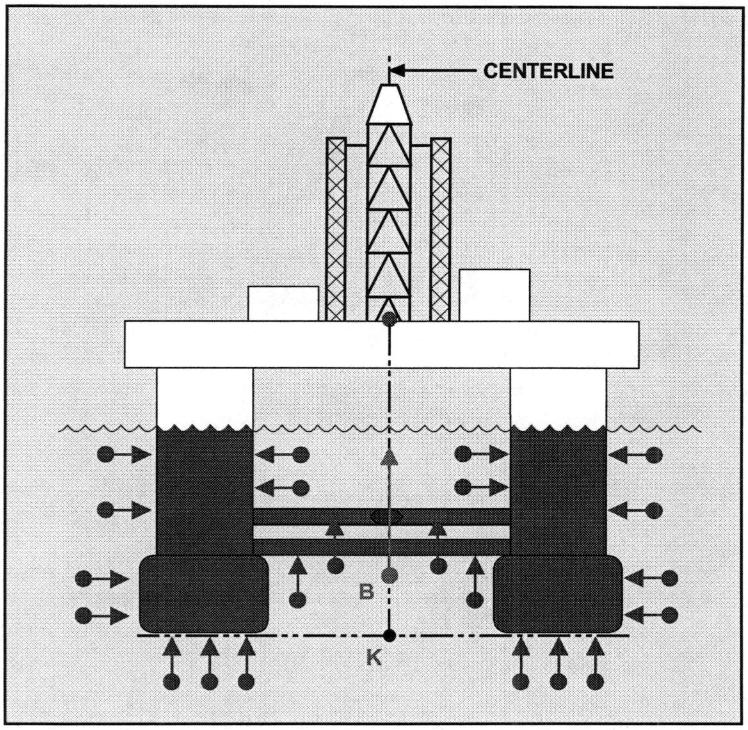

Figure 1.7 The position of the transverse center of buoyancy

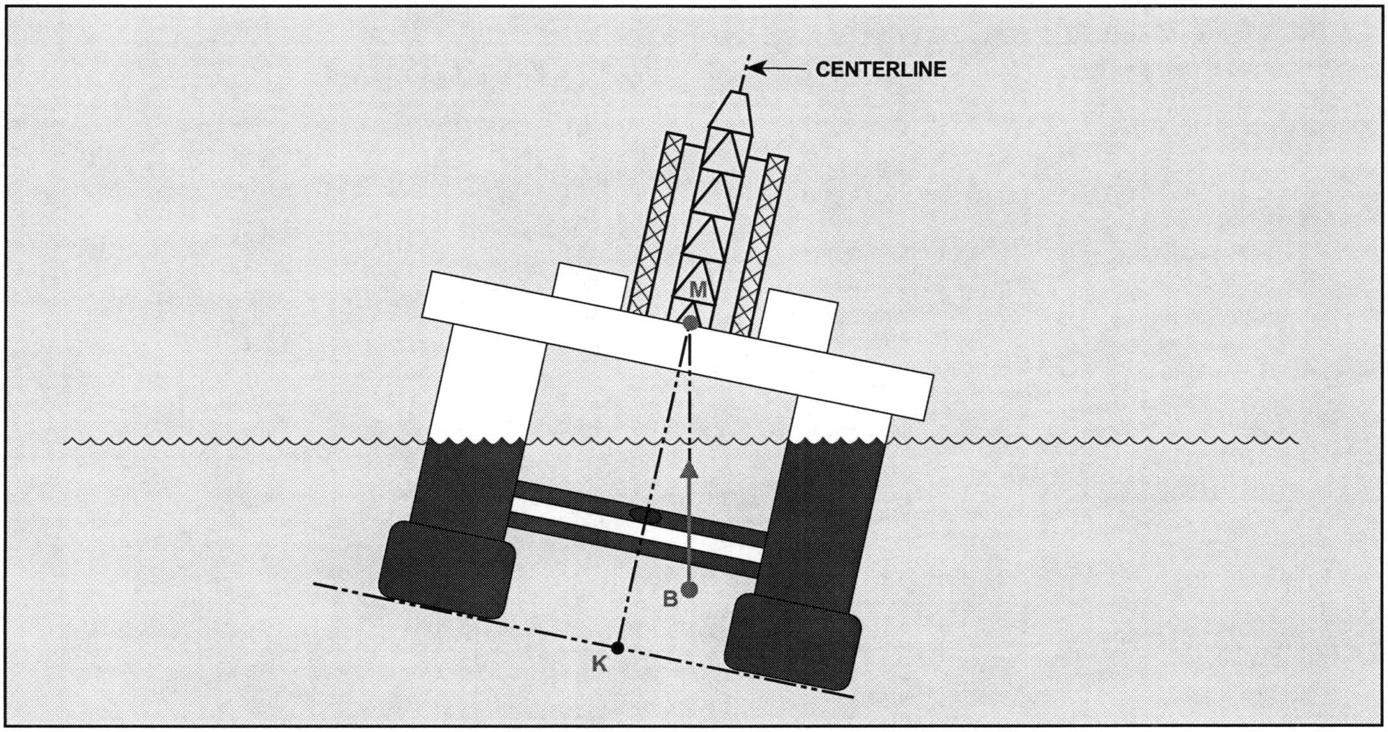

Figure 1.8 Vertical position of B and M on an inclined rig

1.12 THE METACENTER

In the inclined position, the *metacenter* (*M*) is the intersection of the vertical through *B* with the centerline of the rig (see fig 1.8).

1. *M* is considered to be a fixed point for consecutive small angles up to about 8°.
2. The positions of *B* and *M* are related to the draft of the rig. Because the position of *B* is known, the distances B*M* and *KM* can be calculated.

1.13 THE HEIGHT OF THE METACENTER (KM)

The *height of the metacenter* (*KM*) is the vertical distance between the keel and the metacenter. The naval architect calculates this distance *KM* for a range of draft intervals for the transverse and longitudinal as KM_T and KM_L. The results are tabulated in the hydrostatic properties of the rig.

1.14 THE METACENTRIC HEIGHT (GM)

The *metacentric height* (*GM*) is the distance between *G* and *M* (in m or ft) (fig. 1.9). Metacentric height should not be confused with the height of the metacenter which is the distance *KM*.

Figure 1.9 shows the transverse positions of *M*, *G*, *B*, and *K* on the centerline with the rig in upright position.

1. *KG* or *VCG* is the vertical distance (m or ft) between *K* and *G*. The weight of the rig (*W*) acts through *G* vertically downward.
2. *KB* or *VCB* is the vertical distance (m or ft) between *K* and *B*. The buoyancy force (equal to *W*) acts vertically and upward.
3. *KM* (the height of the metacenter) is the vertical distance (m or ft) between *K* and *M*. *M* is a fictitious point but is essential to validate the amount of stability.
4. *BM* is the vertical distance (m or ft) between *B* and *M*.
5. *GM* (the metacentric height) is the vertical distance (m or ft) between *G* and *M*.

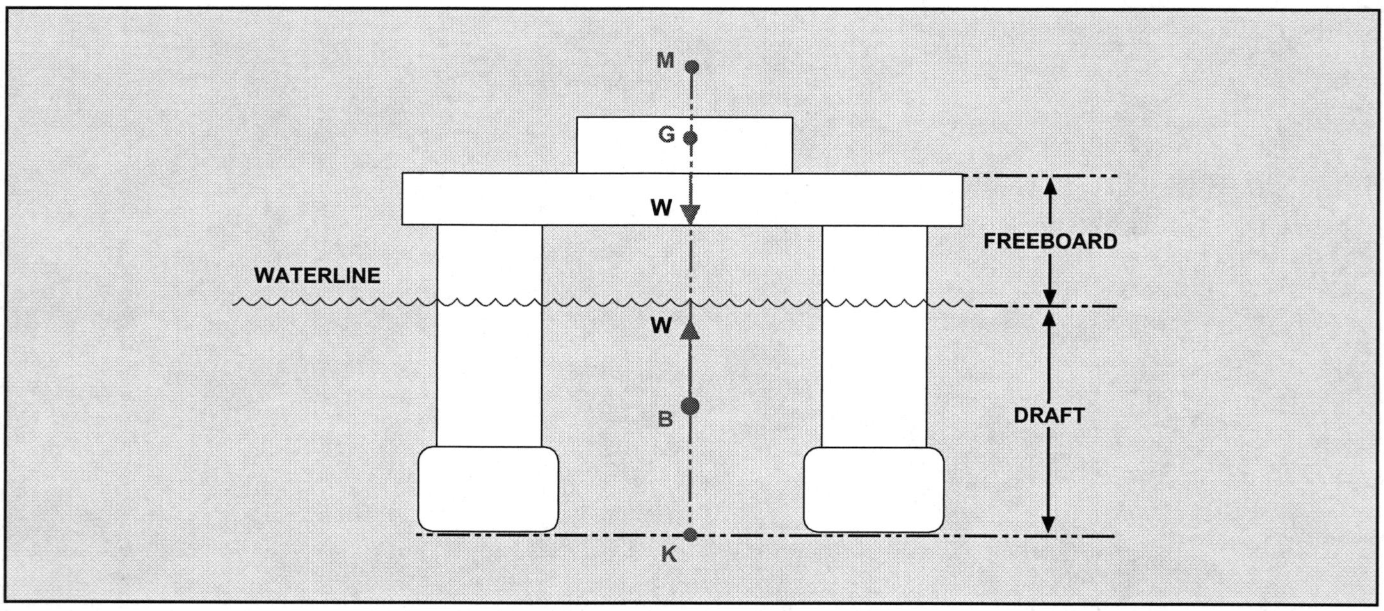

Figure 1.9 Position of M, G, B, and K

1.15 COUPLE

A *couple* (fig. 1.10) is the combination of a pair of equal and opposite parallel forces acting on a body. A couple has the tendency to produce rotation. The magnitude or moment of a couple is one of the forces multiplied times the distance between the parallel forces.

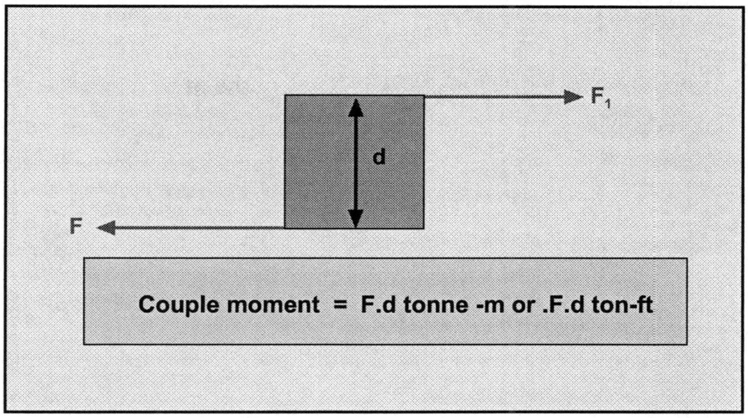

Fig. 1.10 A couple

1.16 THE STABILITY COUPLE

If W is the weight of the displacement of the rig, the inclination caused by the overturning force of wind or waves shown in Figure 1.11 has the following effect:

1. Because the underwater shape changed, the center of buoyancy moves from B_1 to B_2. The buoyancy force through B is vertically upward.
2. G will remain in place because there is no change in the weight distribution. The weight force through G is vertically downward.
3. A couple with the two equal and opposite parallel forces W and an arm GZ forms a restoring force. Whenever the overturning force is removed, the rig will return in this case to the upright position.

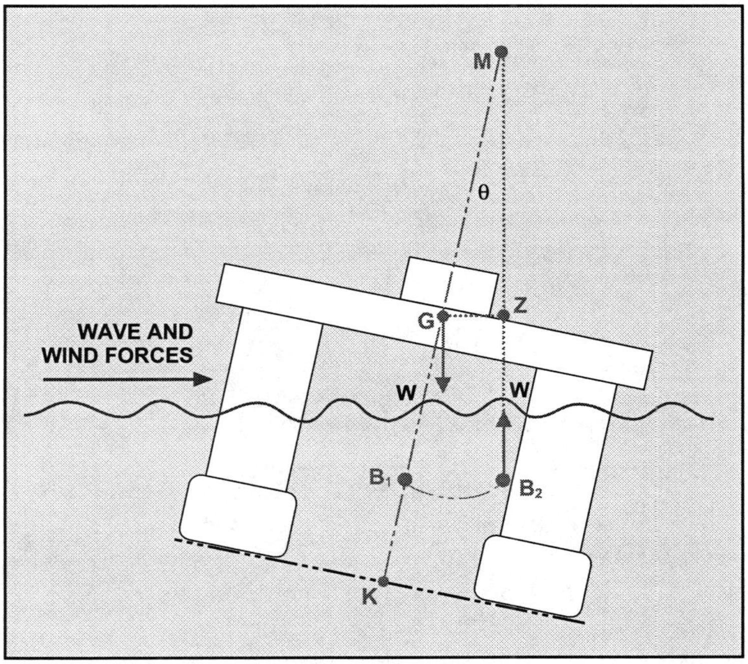

Figure 1.11 The stability couple

1.17 THE RIGHTING ARM

The righting arm (or lever) GZ of stability in Figures 1.11 and 1.12 is the horizontal line through G and the intersection with the vertical through B. The arm of the stability couple GZ is a deciding factor for stability response for any type of vessel.

GZ can be expressed as a function of GM in the triangle MGZ. If θ is the angle, the formula for arm of stability is:

$$GZ = GM \sin \theta$$

1.18 THE RIGHTING MOMENT

The righting moment of the stability couple is the product weight times the arm (figs. 1.11 and 1.12). If *W* is the weight of the displacement of the rig and *GZ* the arm, the formula for the righting moment is:

$$\text{Righting Moment} = W \times GZ \text{ or}$$
$$\text{Righting Moment} = W \times GM \sin \theta$$

The theory of the initial stability with the corresponding couple and moment for small angles is called the *statical stability* and the *moment of statical stability*.

The righting moment formula shows that:

1. The result of a large *GM* value is a large stability couple. The rig is more stable. The rig produces a large restoring force if heeled by external forces.
2. The result of a small *GM* is a small stability couple and a small restoring force. The rig is less stable to counteract the heel caused by the external forces.
3. The interrelated positions of *B* and *M* at a specific draft are fixed.
4. The position of *G* at the same specific draft depends on the weight distribution.
5. Weight stored near the bottom of the hull decreases the value of *KG* and increases the value of *GM*. If weights are moved up, the value of *KG* increases and *GM* decreases.

1.19 STABLE EQUILIBRIUM

Stable equilibrium (fig.1.12) is the tendency of a rig to return to its initial upright position after an external force that heeled the rig to one side is removed. This is called, in general, stability because the restoring force of the stability couple acts in a positive way.

GM = positive

The couple moment = positive

The amount of positive moment depends on:

1. The distance GM.
2. The angle of inclination.
3. The displacement.

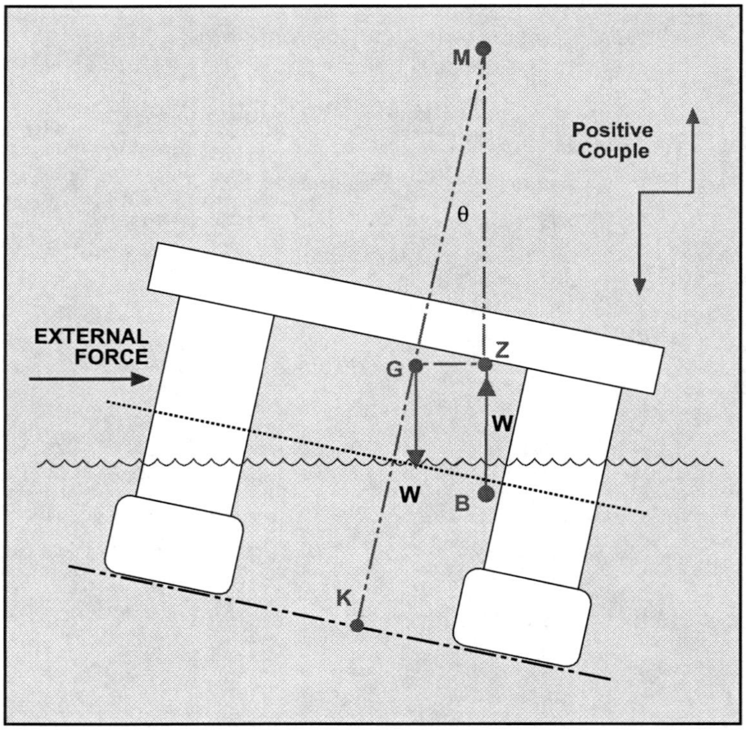

Figure 1.12 Stable equilibrium—positive couple

1.20 NEUTRAL EQUILIBRIUM

If *G* rises to the position of *M*, the new situation is called neutral equilibrium (fig. 1.13).

Neutral equilibrium is the condition when the righting arm is zero and consequently the righting couple is zero. *G* is at the same level as *M*.

GM = zero
The couple moment = 0.
The two opposite forces are now in line.
Without any external forces, the rig lingers around a small angle.
This is called the angle of loll.

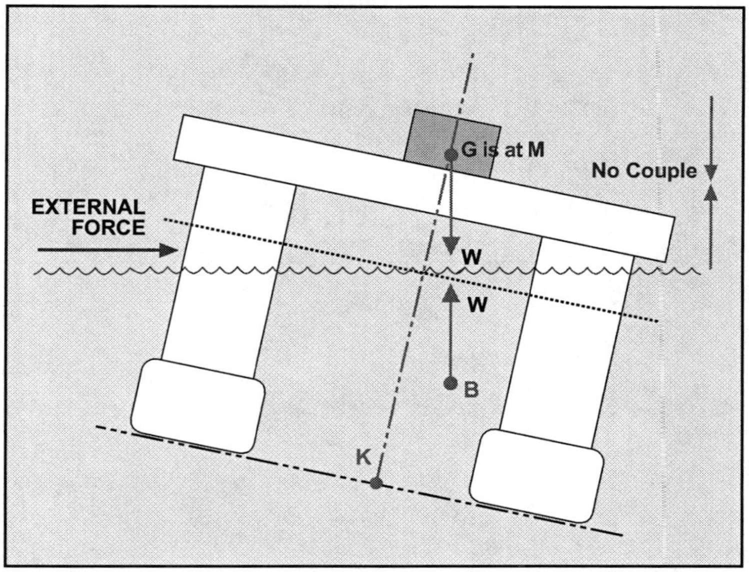

Figure 1.13 Neutral equilibrium

1.21 UNSTABLE EQUILIBRIUM

If more weight is shifted symmetrically upward, *G* will rise above *M*. With *G* still on the centerline, the couple and moment are now working in the same direction as the external force. There is no restoring force.

Unstable equilibrium (fig. 1.14) **is the condition where the rig continues to incline after an external force caused an initial small angle.**

GM = negative
The couple moment = negative
The rig may capsize.

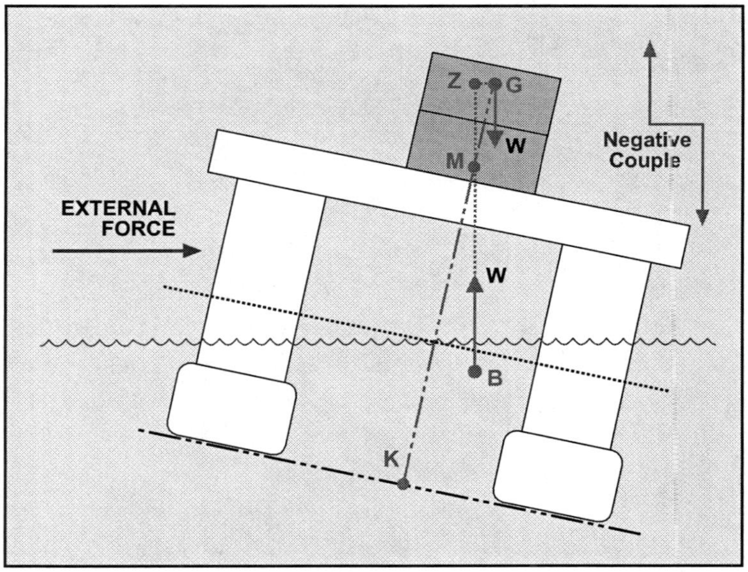

Figure 1.14 Unstable equilibrium—negative couple

1.22 STIFF AND TENDER

A rig with a large stability couple will have quick and unpleasant movements. The acceleration is fast with short and jerky movements. This is called a stiff rig or vessel. A rig with a small stability couple will move lazy and slow. This is called a tender rig or vessel.

1.23 TRANSVERSE VERSUS LONGITUDINAL STABILITY

Most rigs are built symmetrically along the transverse or longitudinal line. However, the length and width are not the same. The resistance to rotate along the shorter transverse waterplane is less than the resistance along the longer longitudinal waterplane. Consequently, the distance KM for longitudinal and transverse stability is not the same for a vessel that is not built symmetrically along the two axes. Figure 1.15 shows the transverse and longitudinal cross sections of a drilling vessel and a semisubmersible and the corresponding KM heights.

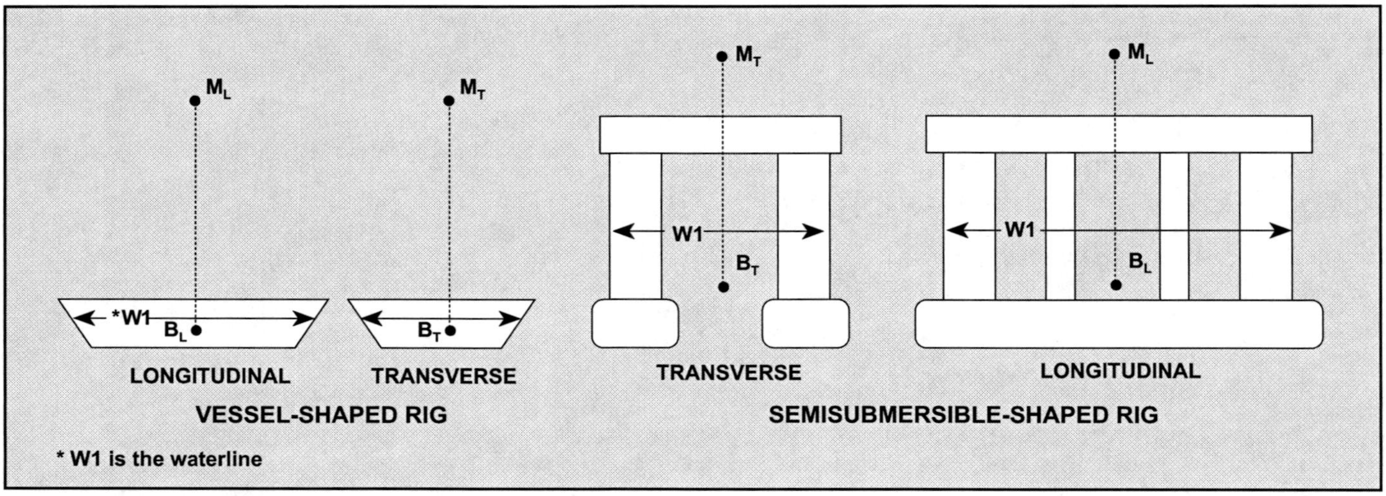

Figure 1.15 Transverse and longitudinal position of M

1.24 CATEGORIES OF STABILITY

Stability can be categorized as:

1. *Intact stability.* Stability under normal conditions without any damage to the rig.
2. *Damaged stability.* The stability of the rig is affected by a damage to the rig hull or structure.
3. *Initial stability for small angles.* The theory and principles for longitudinal and transverse stability for small angles up to 8° including the conditions of:
   a. Stable equilibrium
   b. Neutral equilibrium
   c. Unstable equilibrium
4. *Stability for large angles.* For two conditions:
   a. Static stability
   b. Dynamic stability

Figures 1.16, 1.17, and 1.18 are examples from Rig Type 1's marine operations manual (MOM) with the graphic display of the various points and definitions as discussed in this chapter.

DEFINITIONS AND SYMBOLS
(HULL AND COLUMNS)

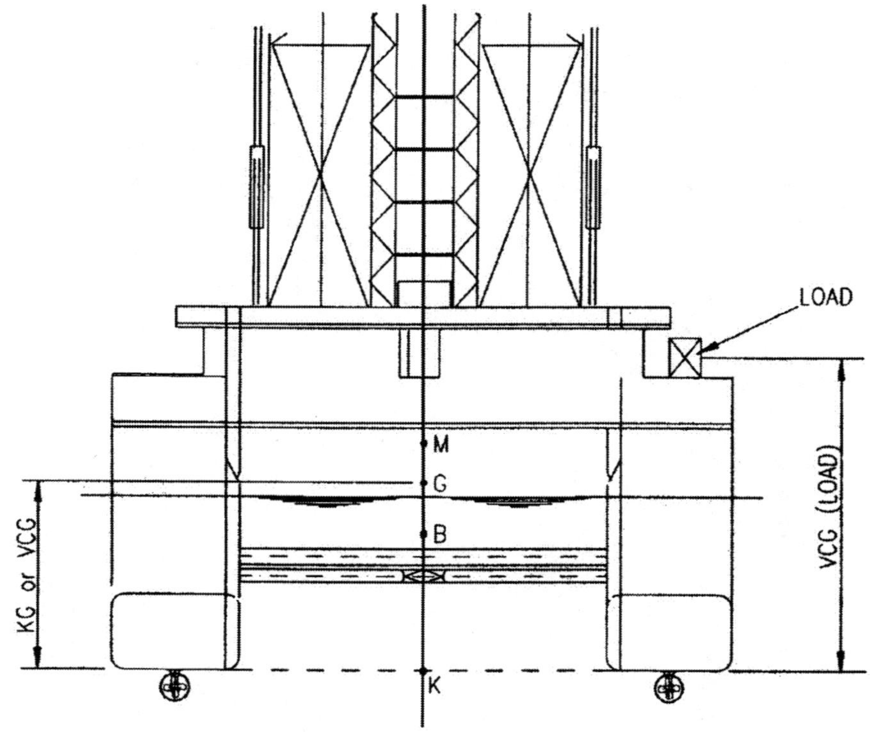

LEGEND:

K	=	Baseline (Keel).
B	=	Center of Buoyancy.
G	=	Center of Gravity of the rig.
M	=	Metacenter.
KM	=	Distance of Metacenter above Baseline. KM_T for Transverse Metacenter and KM_L for Longitudinal Metacenter above Keel.
VCG (LOAD)	=	Vertical Distance of the Center of Gravity of the Load above the Baseline of the Rig.
VM	=	Vertical Moment of the Load = LOAD Weight x VCG of LOAD. This Moment is always Positive.
VCG (Rig)	=	Vertical Center of Gravity of the Rig = Total VMs / Rig' Displacement. (The total vertical moments of all weights onboard [deck loads, tank liquids, stores, drilling loads, etc.] added to the vertical moment of the Rig's Lightship and divided by the Rig's Lightship plus the total of all weights onboard). This Moment is always Positive and **MUST** always be less than the VCG limits given in Table 1.1 and Figure 17.

Figure 1.16 Example of definitions and symbols for Rig Type 1

DEFINITIONS AND SYMBOLS
(PLAN)

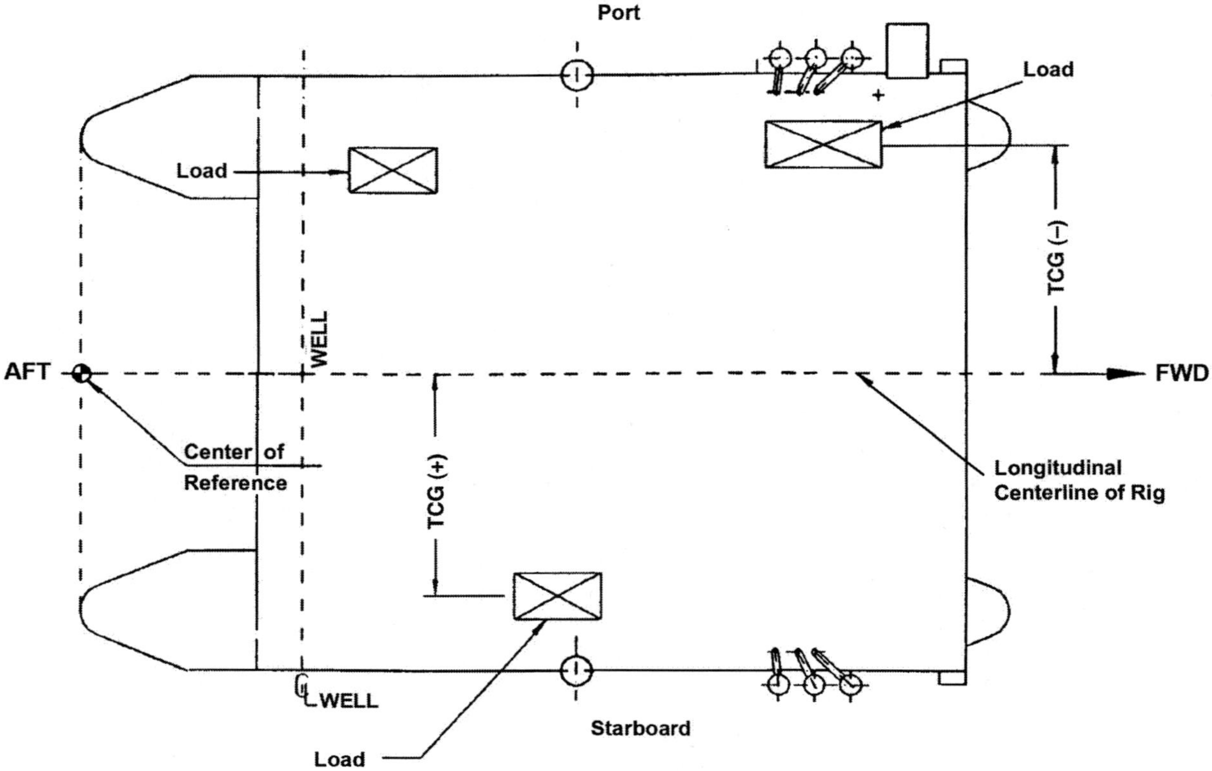

LEGEND:

LCG	=	Longitudinal Distance of a LOAD from Midships (the Transverse Centerline of the Rig). The LCG is Postive (+) Forward of Midships and Negative (–) Aft.
TCG	=	Transverse Distance of a LOAD from the Longitudinal Centerline of the Rig. The TCG is Positive (+) to Starboard and Negative (–) to Port.
LM	=	Longitudinal Moment of a LOAD = Load Weight x LCG of the LOAD. This Moment can be either Positive or Negative.
TM	=	Transverse Moment of a LOAD = Load Weight x TCG of the LOAD. This Moment can be either Positive or Negative.

Figure 1.17 Example of definitions and symbols Rig Type 1

DEFINITIONS AND SYMBOLS
(ELEVATION)

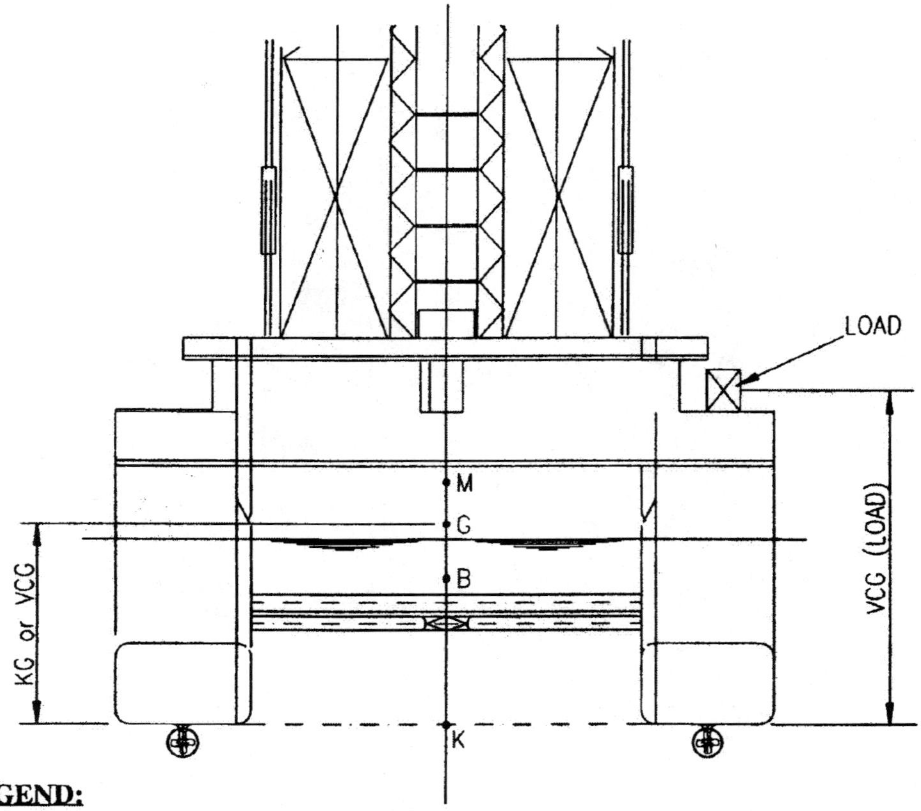

LEGEND:

K	=	**Baseline (Keel).**
B	=	Center of Buoyancy.
G	=	Center of Gravity of the rig.
M	=	Metacenter.
KM	=	Distance of Metacenter above Baseline. KM_T for Transverse Metacenter and KM_L for Longitudinal Metacenter above Keel.
VCG (LOAD)	=	Vertical Distance of the Center of Gravity of the Load above the Baseline of the Rig.
VM	=	Vertical Moment of the Load = LOAD Weight x VCG of LOAD. This Moment is always Positive.
VCG (Rig)	=	Vertical Center of Gravity of the Rig = Total VMs / Rig' Displacement. (The total vertical moments of all weights onboard [deck loads, tank liquids, stores, drilling loads, etc.] added to the vertical moment of the Rig's Lightship and divided by the Rig's Lightship plus the total of all weights onboard). This Moment is always Positive and **MUST** always be less than the VCG limits given in Table 1.1 and Figure 17.

Figure 1.18 Example of definitions and symbols Rig Type 1

1.25 DRAFT

Draft (also spelled as draught) is the vertical distance from the keel to the waterline. The length of thrusters below the hull is normally not included in the draft readings (fig. 1.19).

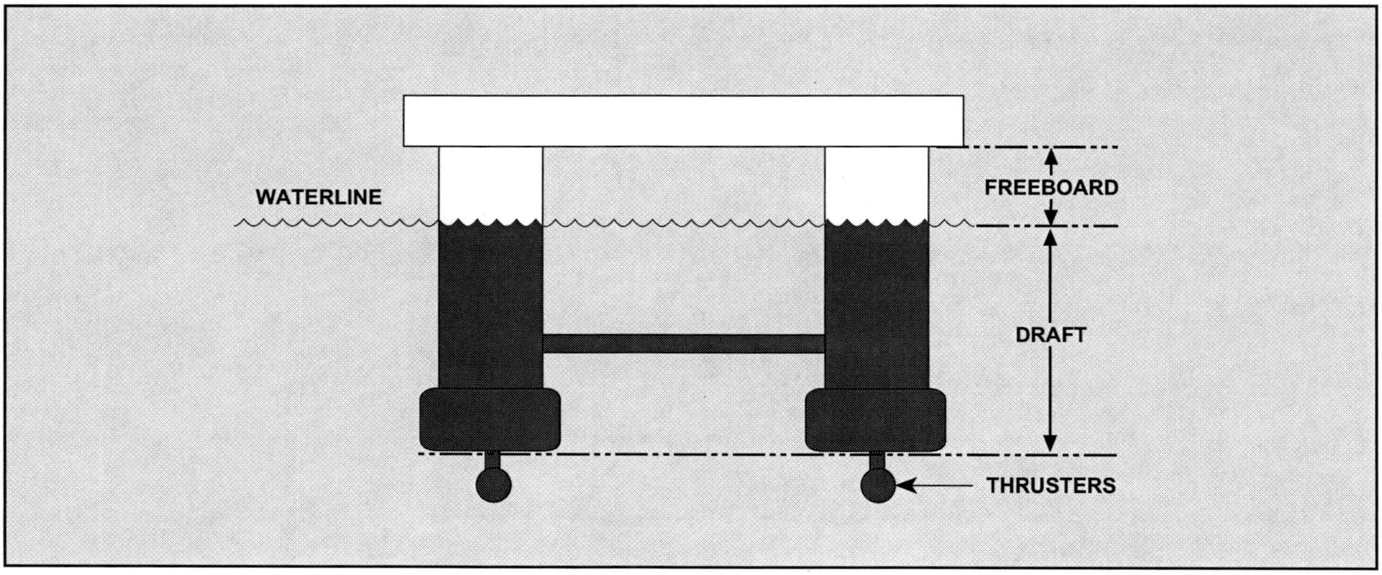

Figure 1.19 Draft

1.26 DRAFT READINGS (Fig. 1.20)

Examples for the metric system.

Examples for the conventional system.

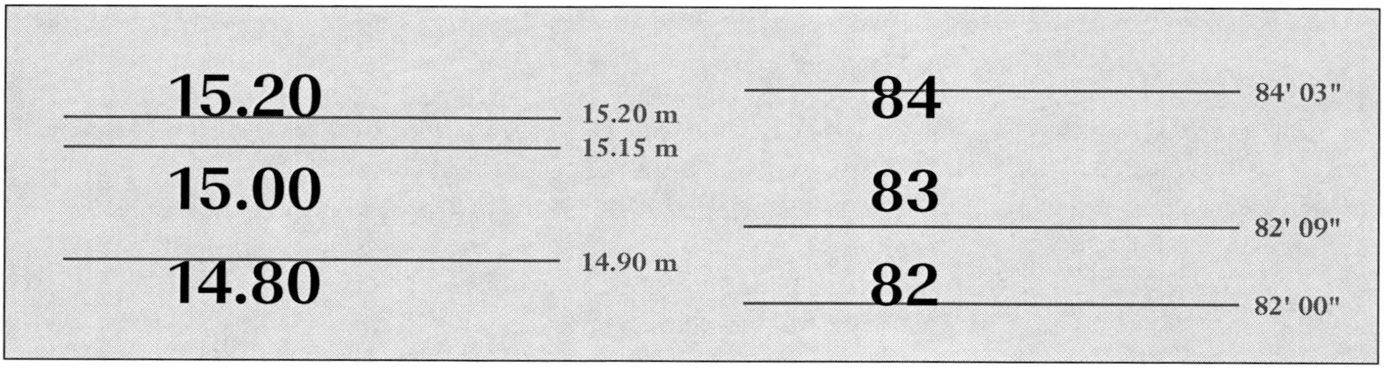

Figure 1.20 Draft marks readings

1.27 TRANSIT DRAFT

For a semisubmersible, this is the draft on the pontoons in a deballasted condition. For a vessel-shaped drilling rig, the transit draft can be any draft.

1.28 OPERATIONAL DRAFT

This is the draft for a semisubmersible in the drilling mode with the best motion characteristics and maximum deck load.

1.29 SURVIVAL DRAFT

This is the draft for a semisubmersible in a standby mode with an adequate and safe air gap during storm conditions.

1.30 MEAN DRAFT

This is the average of the draft readings forward and aft on a vessel-shaped rig or the average of the four column draft readings.

1.31 LAW OF ARCHIMEDES

The *Law of Archimedes* or *Archimedes's Principle* states that a body immersed in a fluid has an apparent loss of weight equal to the weight of the displaced fluid. A body with a weight in air of 30 t and a volume of 20 m^3 will weigh 10 t if immersed in fresh water because the amount of displaced fresh water of 20 m^3 represents a weight of 20 t of upward force.

1.32 FLOATING

Based on the Law of Archimedes, a vessel will *float* if the vessel displaces its own weight of water before it displaces its own volume.

1.33 DISPLACEMENT

Displacement is the weight in tonnes or short tons of the volume of water displaced by a floating body (rig, vessel, barge). Every floating body always displaces its own weight.

Summarizing Archimedes's Law, floating and displacement:

1. A drilling rig will float as long as the weight of the drilling rig does not exceed the weight of the displaced volume of water.
2. The weight of the displaced water is equal to the weight of the drilling rig.
3. The volume of the displacement by the drilling rig at a specific draft is the volume of water displaced by the drilling rig.

1.34 VOLUME OF DISPLACEMENT

Volume of displacement is the volume of liquid displaced by the rig in m^3 or ft^3. It is the underwater volume of the rig.

1.35 LIGHTSHIP DISPLACEMENT

Lightship displacement (also called *light weight*, *lightship weight*, *light displacement*) is the weight of the steel structure of the rig and all fixed equipment but without any ballast fuel, consumables, deck loads, bulk material, stores, drilling tubulars, and loose equipment, etc. The anchors and chain may be part of the lightship weight.

1.36 THE WATERPLANE AREA (WPA)

The *waterplane area* (*WPA*) is the area of a horizontal cross section of the rig at the waterline for any given draft.

1.37 TOTAL DISPLACEMENT (fig. 1.21)

TOTAL DISPLACEMENT = LIGHTSHIP WEIGHT + VARIABLE LOAD

Figure 1.21 shows the breakdown of the total displacement.

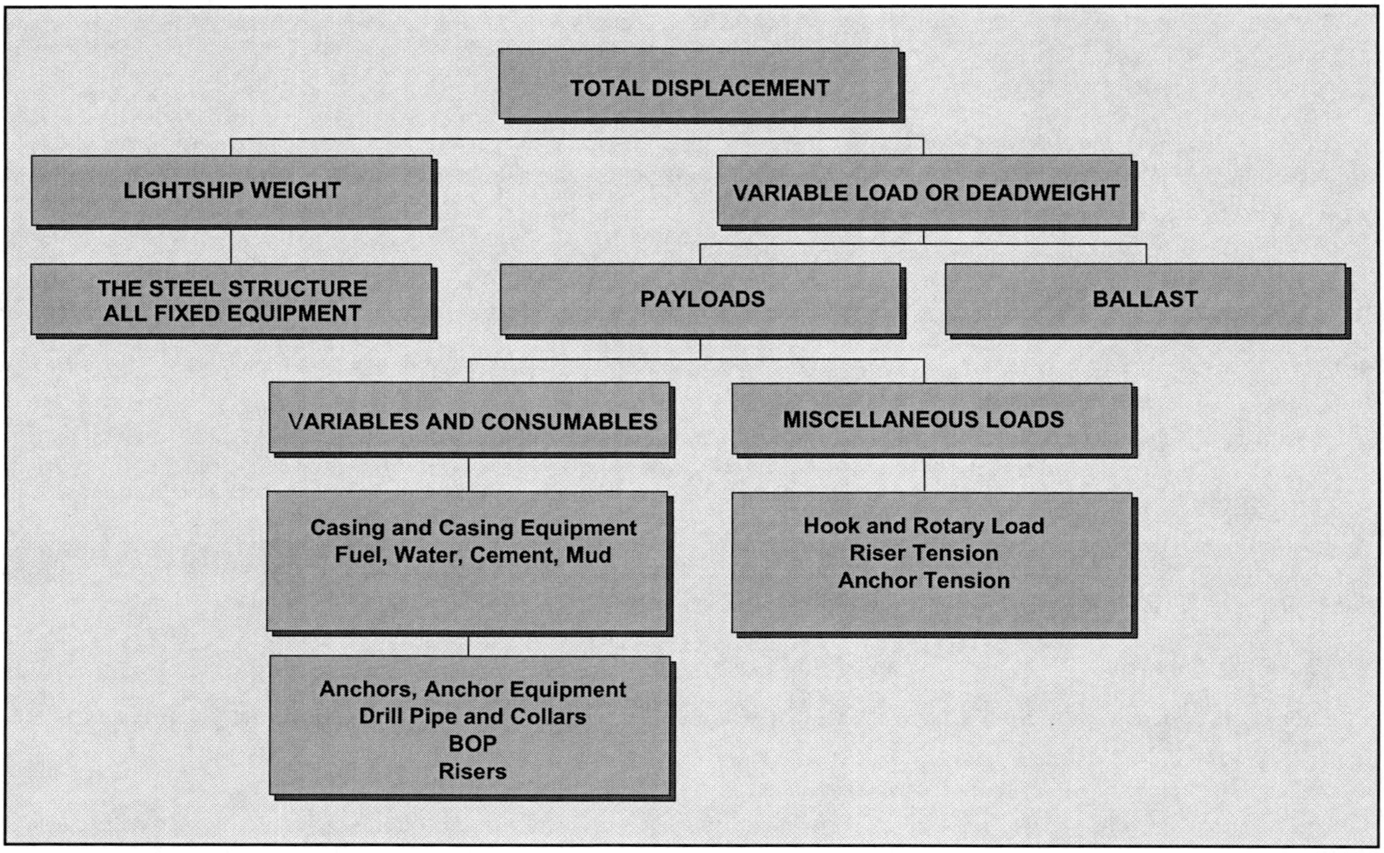

Figure 1.21 **Total displacement**

1.38 FREEBOARD (fig. 1.19)

Freeboard is the vertical distance in m or ft between the waterline and the upper side of the main deck. The main deck is the uppermost deck with watertight integrity.

1.39 TRIM

Trim is the longitudinal inclination of the rig. Trim is the difference between the forward and aft draft readings. A trim of 1'6" aft means that the draft aft is 1'6" deeper than the draft forward.

1.40 HEEL OR LIST

Heel or *list* is the transverse inclination of the rig. For neutral equilibrium, heel or list is referred to as the *angle of loll.*

1.41 CHANGE OF TRIM (COT) AND CHANGE OF HEEL (COH)

Change of trim (*COT*) is the difference of the initial and final trim caused by a load change in the forward aft direction.

Change of heel (*COH*) is the difference of the initial and final heel caused by a load change in the port starboard direction.

1.42 CENTER OF FLOTATION (COF)

Center of flotation (*COF*) is the tipping center of the vessel or rig. It is the point about which the rig heels or trims. It is called the center of gravity of the waterplane. With an asymmetric waterline, the position of the *COF* does not coincide with the intersection of centerlines of the rig. In Figure 1.22, the top view of a jackup shows that the *COF* lies on the longitudinal centerline but aft of the transverse centerline of the rig.

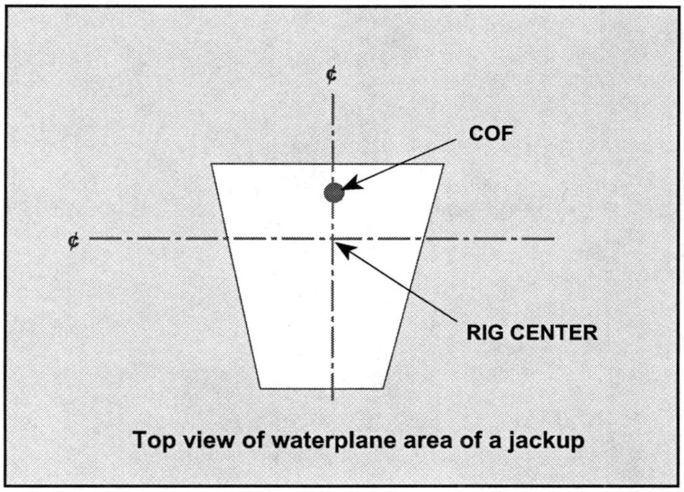

Figure 1.22 Center of flotation

1.43 RESERVE BUOYANCY

Reserve buoyancy is the volume of the watertight enclosed spaces above the waterline. Combined with the freeboard reserve, buoyancy is the safety margin to ensure enough reserve buoyancy is available in case of large angles of heel or trim caused by environmental forces or a damaged stability condition (fig 1.23).

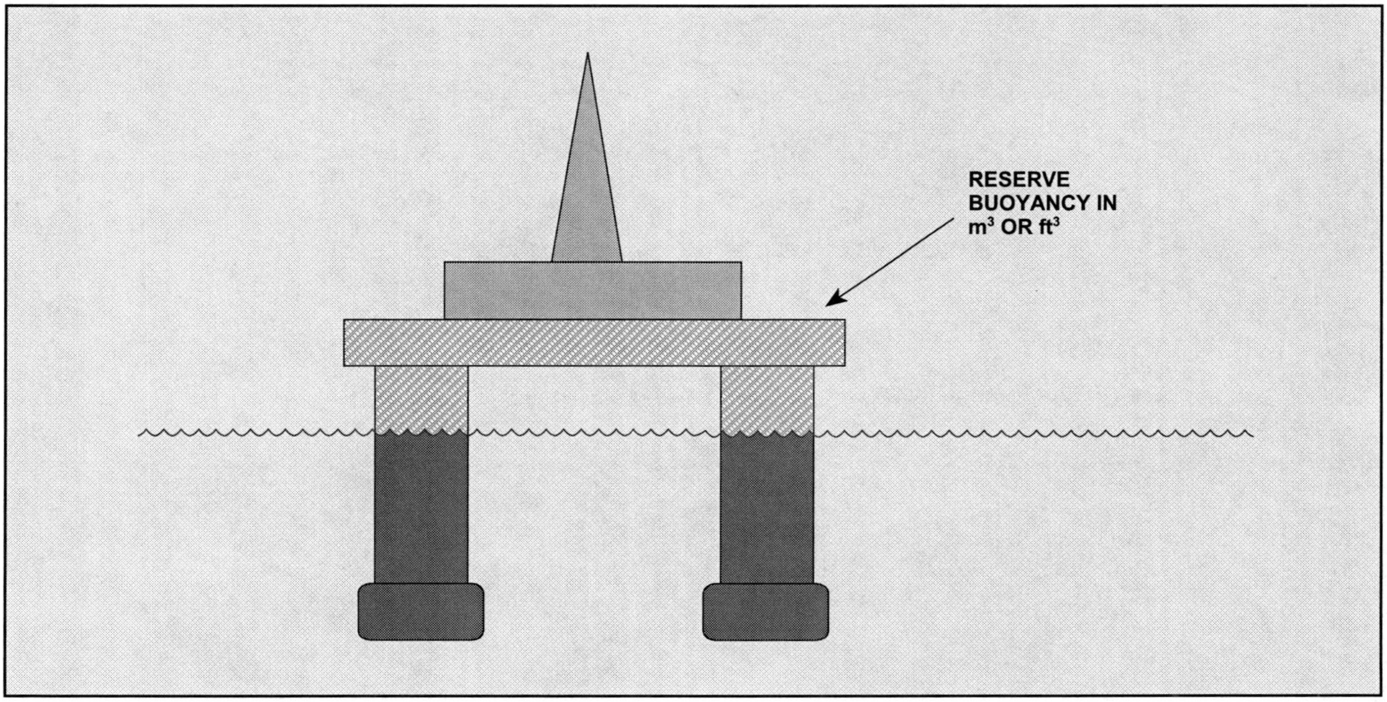

Figure 1.23 Reserve buoyancy

1.44 LOAD LINE OR PLIMSOLL MARK

The *load line* or *Plimsoll mark* (fig. 1.24) displays the maximum allowed draft at any time. A deck line welded on the hull indicates the uppermost watertight deck above the load line. The distance between the deck line and Plimsoll mark is the minimum required freeboard.

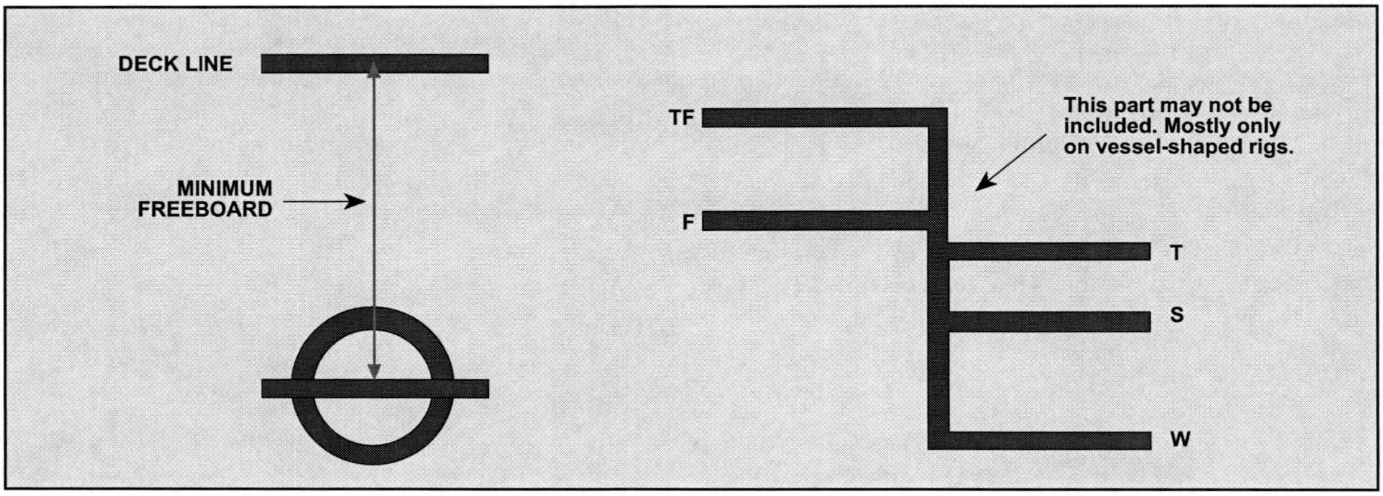

Figure 1.24. The load line or Plimsoll mark

1.45 DENSITY

Density is mass or weight per unit volume.

$$\text{DENSITY} = \frac{\text{MASS}}{\text{VOLUME}} \text{ OR}$$

$$\text{MASS} = \text{DENSITY} \times \text{VOLUME}$$

1.46 RELATIVE DENSITY

Relative density (*RD*) is the ratio of the density of a substance to the density of a standard substance.

$$\text{RELATIVE DENSITY} = \frac{\text{DENSITY OF A SUBSTANCE}}{\text{DENSITY OF FRESH WATER}} \text{ OR}$$

$$\text{RELATIVE DENSITY} = \frac{\text{THE WEIGHT OF ANY GIVEN VOLUME OF SUBSTANCE}}{\text{WEIGHT OF AN EQUAL VOLUME OF FRESH WATER}}$$

Depending on the unit system used on the rig, density and relative density are measured in:

kg/m^3 or lb/ft^3

For our purpose the numbers used in stability calculations for density and relative density are:

FOR THE SI SYSTEM

The density of fresh water is 1,000 kg/m^3 or 1.0 t/m^3

The density of seawater is 1,025 kg/m^3 or 1.025 t/m^3

FOR THE CONVENTIONAL SYSTEM

The density of fresh water is 1,000 oz/ft^3 or 62.5 lb/ft^3

The density of seawater is 1,025 oz/ft^3 or 64.0 lb/ft^3

FOR BOTH SYSTEMS

The relative density of fresh water is 1.0

The relative density of seawater is 1.025

1.47 VOLUME VERSUS WEIGHT

From the previous explanation it is known that weight is equal to volume multiplied by density. In more details:

$$\text{Volume (V)} = \frac{\text{Weight (W)}}{\text{Density } (\delta)}$$

therefore

$$\text{Volume of Displaced Water} = \frac{\text{Weight of Displaced Water}}{\delta \text{ of Water}}$$

or

$$\text{Underwater Volume} = \frac{\text{Weight of the body}}{\delta \text{ of Water}}$$

For the metric system we use:

$$\text{Volume in m}^3 = \frac{\text{Weight in t}}{\delta \text{ in t/m}^3}$$

$$\text{Weight in t} = \text{Volume in m}^3 \times \delta \text{ in t/m}^3$$

The formulas for the conventional system are more complicated:

$$\text{For Fresh Water: V in ft}^3 = \frac{\text{W in st} \times 2{,}000 \text{ lb} \times 16 \text{ oz}}{1{,}000 \text{ oz/ft}^3} = \text{W in st} \times 32 \text{ ft}^3$$

$$\text{For Seawater: V in ft}^3 = \frac{\text{W in st} \times 2{,}000 \text{ lb} \times 16 \text{ oz}}{1{,}025 \text{ oz/ft}^3} = \text{W in st} \times 31.2 \text{ ft}^3$$

The formulas for the conventional system are derived from the units:

1 lb	=	16 oz
1 st	=	2,000 lb
Density of fresh water	=	1,000 oz/ft^3
Density of seawater	=	1,025 oz/ft^3

The formulas show that:

One (1) st of fresh water = 32 ft^3 and one (1) st of seawater = 31.2 ft^3

The following formulas are used to calculate the volume in ft^3 occupied by a given weight in tonnes (metric) of fresh water or seawater.

$$\text{For Fresh Water: V in ft}^3 = \frac{\text{W in t} \times 2{,}205 \text{ lb} \times 16 \text{ oz}}{1{,}000 \text{ oz/ft}^3} = \text{W in t} \times 35.3 \text{ ft}^3$$

$$\text{For Seawater: V in ft}^3 = \frac{\text{W in t} \times 2{,}205 \text{ lb} \times 16 \text{ oz}}{1{,}025 \text{ oz/ft}^3} = \text{W in t} \times 34.4 \text{ ft}^3$$

1.48 TONNES PER CENTIMETRE (TPC) IMMERSION—TONS PER INCH IMMERSION (TPI)

The *tonnes per centimetre* (*TPC*) immersion is the weight in tonnes which must be loaded/discharged to change the average draft of the rig in seawater by one cm. The *tons per inch* (*TPI*) immersion is the weight in short tons which must be loaded/discharged to change the average draft of the rig in seawater by one inch.

The value of TPC or TPI depends on the size of the *waterplane area* (*WPA*) (fig. 1.25).

Figure 1.25 shows one side of the pontoon and columns of a semisubmersible. The figure clearly shows that the volume of the waterplane with a 1 cm or 1" thickness at pontoon draft is much larger than the waterplane with the same thickness at a draft around columns. To change from *WL1* to *WL2* on the pontoons required much more weight than to change the same amount at column draft.

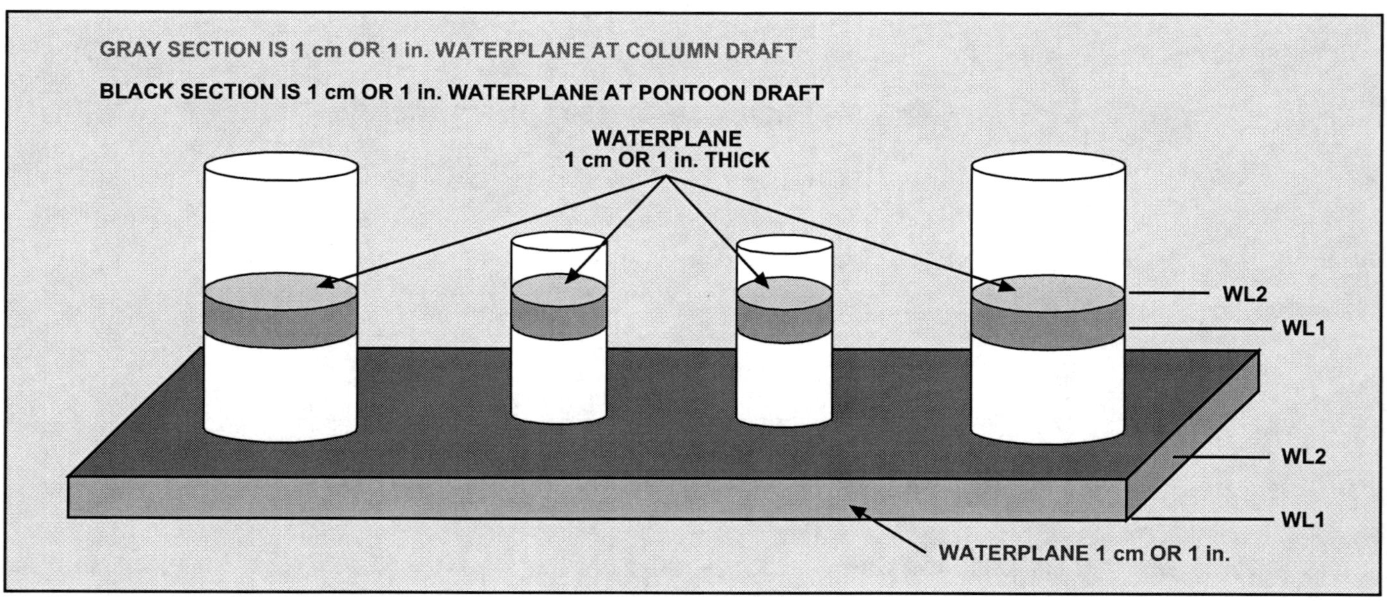

Figure 1.25 Waterplanes

1.49 CALCULATION OF TPC OR TPI

Calculation for TPC for seawater.

If WPA is the waterplane area in m^2, the volume (V) of the 1 cm layer is $\frac{WPA}{100}$ m^3.

To submerge the rig 1 cm, it is necessary to add a weight on the rig which is equal to the weight of the displaced water layer of one cm.

$$\text{Weight} = V \times \delta$$

$$\delta\ \text{Seawater} = 1.025\ t/m^3$$

$$\text{Substitute V (with 1 cm thickness)} = \frac{WPA}{100} \text{ to find the weight of TPC}$$

$$TPC_{sw} = \frac{\delta \times WPA}{100}$$

$$TPC_{sw} = \frac{1.025\ WPA}{100}$$

Calculation for TPI for seawater.

The calculation of TPI is based on the same principles as the calculation of TPC.

$$\text{The Volume (V) for 1 in. sinkage in ft}^3 = \frac{WPA\ (ft^2)}{12}$$

$$TPI_{sw} = \frac{WPA\ (ft^2) \times 1{,}025}{12 \times 16 \times 2{,}000} \quad \text{so}$$

$$TPI_{sw} = \frac{WPA}{375}$$

The naval architect calculates TPC and TPI for each cm or inch of change in draft. The results are printed in the hydrostatic tables. See Table 1.2.

TPC and TPI are the same for drafts with the same waterplane. If the waterplane changes, TPC and TPI values change. This is very obvious with the semisubmersible configuration when the pontoons and transverse tubulars start to submerge.

1.50 HYDROSTATIC PROPERTIES OR HYDROSTATIC TABLES

Hydrostatic properties or *tables* are tables (specific for each rig) with precalculated information. The naval architect obtains this information during the design period of the rig. The information displays properties against a range of draft readings such as displacement, waterplane area, ton per cm, and the positions of important points required for the stability calculations. Table 1.2 is an example of the hydrostatic properties of Rig Type 1.

TABLE 1.2

Example of Rig Type 1 Hydrostatic Properties

HYDROSTATIC PROPERTIES

Metric - (Sheet 3 of 8)

Draft (m)	Displt. (tonnes)	VCB (m)	LCB (m)	TCB (m)	WPA (m^2)	TPC (t/cm)	LCF (m)	TCF (m)	KM_L (m)	KM_T (m)	MCT (6) (t-m/cm)	MCH (6) (t-m/cm)
7.00	19,548	3.49	53.26	0.00	2,727	28.09	53.25	0.00	119.45	114.34	219	324
7.10	19,829	3.54	53.26	0.00	2,727	28.09	53.25	0.00	117.86	112.82	219	324
7.20	20,110	3.59	53.26	0.00	2,727	28.09	53.25	0.00	116.31	111.34	220	325
7.30	20,391	3.64	53.26	0.00	2,727	28.09	53.25	0.00	114.81	109.91	220	325
7.40	20,672	3.69	53.26	0.00	2,727	28.09	53.25	0.00	113.35	108.52	220	325
7.50	20,952	3.74	53.26	0.00	2,727	28.09	53.25	0.00	111.93	107.16	220	325
7.60	21,233	3.79	53.26	0.00	2,727	28.09	53.25	0.00	110.55	105.84	220	326
7.70	21,514	3.84	53.26	0.00	2,727	28.09	53.25	0.00	109.20	104.56	221	326
7.80	21,795	3.89	53.26	0.00	2,727	28.09	53.25	0.00	107.90	103.31	221	326
7.90	22,076	3.94	53.26	0.00	2,727	28.09	53.25	0.00	106.62	102.10	221	327
8.00	22,357	3.99	53.26	0.00	2,727	28.09	53.25	0.00	105.38	100.92	221	327
8.10	22,637	4.04	53.26	0.00	2,712	27.93	53.23	0.00	104.08	99.20	221	325
8.20	22,916	4.09	53.26	0.00	2,696	27.77	53.21	0.00	102.82	97.52	221	324
8.30	23,193	4.14	53.26	0.00	2,680	27.60	53.20	0.00	101.59	95.88	221	322
8.40	23,468	4.19	53.26	0.00	2,664	27.44	53.18	0.00	100.39	94.28	221	321
8.50	23,741	4.24	53.25	0.00	2,647	27.27	53.16	0.00	99.23	92.72	221	319
8.60	24,013	4.29	53.25	0.00	2,631	27.09	53.14	0.00	98.09	91.20	221	317
8.70	24,283	4.34	53.25	0.00	2,614	26.92	53.12	0.00	96.98	89.71	221	316
8.80	24,551	4.39	53.25	0.00	2,597	26.74	53.11	0.00	95.90	88.26	221	314
8.90	24,818	4.43	53.25	0.00	2,579	26.57	53.09	0.00	94.85	86.83	221	312
9.00	25,083	4.48	53.25	0.00	2,183	22.49	51.47	0.03	85.77	72.65	202	264
9.10	25,237	4.51	53.23	0.00	1,494	15.39	51.00	0.00	85.45	51.49	202	188
9.20	25,391	4.54	53.22	0.00	1,492	15.37	51.03	0.00	84.89	51.18	202	188
9.30	25,544	4.57	53.21	0.00	1,490	15.35	51.05	0.00	84.32	50.86	202	188
9.40	25,697	4.59	53.19	0.00	1,488	15.32	51.08	0.00	83.72	50.53	202	188
9.50	25,851	4.62	53.18	0.00	1,486	15.30	51.10	0.00	83.11	50.19	202	188
9.60	26,004	4.65	53.17	0.00	1,484	15.29	51.13	0.00	82.48	49.85	201	188
9.70	26,156	4.68	53.16	0.00	1,483	15.27	51.16	0.00	81.83	49.50	201	188
9.80	26,309	4.71	53.15	0.00	1,480	15.24	51.18	0.00	81.17	49.14	201	187
9.90	26,461	4.74	53.13	0.00	1,476	15.20	51.20	0.00	80.49	48.77	200	187
10.00	26,613	4.77	53.12	0.00	1,184	12.20	53.38	0.39	56.98	39.77	142	153
10.10	26,712	4.79	53.12	0.00	954	9.83	52.68	0.00	39.28	32.94	99	128
10.20	26,809	4.81	53.12	0.00	934	9.62	52.63	0.00	37.95	32.23	96	125
10.30	26,905	4.83	53.12	0.00	913	9.41	52.58	0.00	36.64	31.51	93	123
10.40	26,999	4.85	53.12	0.00	933	9.61	52.68	0.00	35.67	31.05	90	122
10.50	27,097	4.87	53.12	0.00	968	9.97	52.83	0.00	34.85	30.70	89	121
10.60	27,199	4.89	53.11	0.00	997	10.26	52.94	0.00	34.01	30.31	87	119
10.70	27,301	4.91	53.11	0.00	989	10.18	52.94	0.00	32.92	29.70	84	118
10.80	27,402	4.93	53.11	0.00	977	10.06	52.93	0.00	31.83	29.06	82	115
10.90	27,502	4.95	53.11	0.00	965	9.94	52.92	0.00	30.76	28.42	79	113

1.51 CHANGE OF DRAFT GOING FROM SEAWATER TO FRESH WATER

When moving between fresh water and seawater, the draft of the rig will change. The change in draft is caused by the difference in density of the liquid in which the rig floats.

The density of salt water is 1,025 kg/m³ or 1,025 oz/ft³ (64 lb/ft³). The density of fresh water is 1,000 kg/m³ or 1000 oz/ft³ (62.5 lb/ft³).

For a box-shaped vessel (metric and Imperial systems) the new draft can be calculated as follows:

$$\text{new volume} \times \text{new density} = \text{old volume} \times \text{old density} \rightarrow \frac{\text{new volume}}{\text{old volume}} = \frac{\text{old density}}{\text{new density}}$$

For a box-shaped vessel: $\frac{L \times B \times \text{new draft}}{L \times B \times \text{old draft}} = \frac{\text{old } \delta}{\text{new } \delta}$ or

$$\frac{\text{new draft}}{\text{old draft}} = \frac{\text{old } \delta}{\text{new } \delta}$$

If the rig must remain at the same draft when moving to water with a different δ, the formula is:

$$\frac{\text{new displacement}}{\text{new } \delta} = \frac{\text{old displacement}}{\text{old } \delta}$$

$$\text{new displacement} = \text{old displacement} \ \frac{\text{new } \delta}{\text{old } \delta}$$

The formula $\frac{\text{new draft}}{\text{old draft}} = \frac{\text{old } \delta}{\text{new } \delta}$ is only correct for box-shaped vessels.

The change of draft (COD) for other shaped vessels for the metric system can be calculated:

The displacement (W) remains the same if the density changes from δ to δ_1.

The change in volume can be calculated.

$$\text{Volume}_{sw} = \frac{W}{\delta} \text{ and the new density Volume } \delta_1 = \frac{W}{\delta_1}$$

The difference between the two volumes represents the increase in draft.

$$\text{Increase in volume } \Delta \text{ Volume} = \frac{W}{\delta_1} - \frac{W}{\delta} \text{ or } \Delta \text{ Volume} = \frac{W\delta}{\delta_1\delta} - \frac{W\delta}{\delta\delta_1} = \frac{W\delta - W\delta_1}{\delta\delta_1}$$

$$\therefore \Delta \text{ Volume} = \frac{W}{\delta\delta_1}(\delta - \delta_1) \quad \text{Formula 1}$$

$$\text{But also } \Delta \text{ Volume} = \frac{\text{COD} \cdot \text{TPC}_{sw}}{\delta} \quad \text{Formula 2}$$

$$\text{Substitute Formulas 1 and 2} \rightarrow \text{COD} = \frac{W}{\text{TPC}_{sw}} \cdot \frac{\delta - \delta_1}{\delta_1}$$

The calculation of the COD for the Imperial system is:

$$V(\text{ft}^3)\ SW = \frac{W(\text{s.t.}) \cdot 2{,}000 \times 16}{1{,}025} = W(\text{s.t.}) \cdot 31.2 \text{ ft}^3 \text{ and } V(\text{ft}^3)\ FW = \frac{W(\text{s.t.}) \cdot 2{,}000 \times 16}{1{,}000} = W(\text{s.t.}) \cdot 32 \text{ ft}^3$$

$$\text{Change of Volume} = 32.0W - 31.2W$$

$$\text{COD in inch for s.t.} = \frac{\text{Change in V (ft}^3) \cdot 12}{\text{WPA in ft}^2} \text{ and because } \text{TPI}_{sw} = \frac{\text{WPA}}{375} \rightarrow \text{WPA} = 375\ \text{TPI}_{sw}$$

$$\text{or COD in inch for s.t.} = \frac{0.8 \times W \times 12}{375\ \text{TPI}_{sw}} \text{ or COD} = \frac{W}{40\ \text{TPI}_{sw}}$$

EXERCISE 2
Chapter 1

1. A wooden block is 9.2 m long, 1.6 m wide, and 0.7 m high. The weight is 1,900 kg. What is the density of the wood?

2. A floating body displaces 55 ft^3 in salt water. What is the weight of the body in lb?

3. A ship displaces 5,000 m^3 of water with a density of 1,025 kg/m^3. Find the weight in metric tonnes.

4. A drilling vessel displaces 12,400 st. What is the volume of displacement in ft^3 in fresh water?

5. The density of a body is 300 kg/m^3. Calculate the relative density of the body.

6. What is the volume in ft^3 in seawater of a rig with a weight of 22,500 metric tonnes?

7. A semisubmersible floats in seawater and displaces 598.000 ft^3. What is the displacement in short tons?

8. A box-shaped barge is 320 ft long by 37 ft wide with a weight of 2,750 st. Find the change in mean draft on passing from seawater into fresh water.

9. A rectangular shaped barge is 45 m long and 9.1 m wide. Calculate the draft in seawater if the weight is 555 mt.

10. If the draft of a semisubmersible in salt water (δ = 1,025 kg/m^3) is 18 m. The displacement is 33,000 mt and TPC = 7.04. What will the draft be in water with a δ = 1,015 kg/m^3?

11. The displacement of a rig is 13,500 mt with a draft of 8.5 m in salt water. How much cargo must be removed to maintain the same draft in water with a relative density of 1.002?

12. What is the waterplane area of a box-shaped barge 250 m long, 35 m wide, and 15 m in depth? What is the TPC for this barge for salt water?

13. What are the drafts for the two examples?

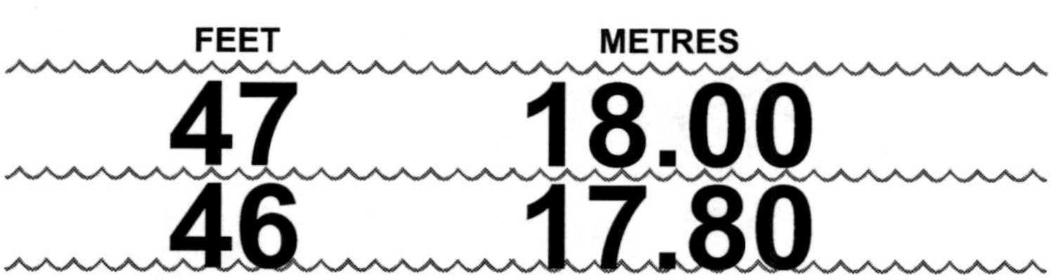

14. A boxed shaped rig is 445 ft long, 72 ft wide, and the main deck is at 27 ft above keel. The lightship weight is 13,000 st.
    a. What is the mean draft and freeboard in salt water for lightship?
    b. After sea trial, 5,500 st is loaded. Calculate the new draft and freeboard.
    c. What is the reserve buoyancy in ft^3 in loaded condition?
    d. What is the TPI for seawater?
    e. What will the draft be in loaded condition in water with δ = 1,008 oz/ft^3?
    f. What is the displacement in ft^3 in water with δ = 1,008 oz/ft^3?

1.52 PERMEABILITY

Permeability with respect to stability is the percentage of water that can fill an enclosed room compared to the total volume of the room. The permeability of a void space without any equipment is 100%. A storeroom filled with equipment and stores may only have a permeability of 55%. In case of a flooded compartment, permeability is an important factor to calculate the amount of possible water ingress.

1.53 VARIABLE LOAD OR DEADWEIGHT (fig. 1.21)

Variable load or *deadweight* is the difference between the lightship weight displacement and the displacement at a loaded draft.

1.54 MEAN DRAFT (MD)

Mean draft (*MD*) is the draft amidships at the intersection of the transverse and longitudinal centerlines.

$$\text{Mean draft drilling vessel} = \frac{\text{Draft reading forward} + \text{Draft reading aft}}{2}$$

$$\text{Mean draft semisubmersible} = \frac{\text{The sum of the draft readings of four columns}}{4}$$

1.55 TRUE MEAN DRAFT (TMD)

The *true mean draft* (*TMD*) is the draft at the *center of flotation* (*F*). The position of *F* depends on the waterplane area. If the position *F* is not amidships, the *TMD* and *MD* are different.

The position of *F* in Figure 1.26 shows the situation where the *longitudinal center of flotation* (*LCF*) does not coincide with amidships because of an asymmetric waterplane area. The *TMD* in this case is less than the *MD*. The correction is the distance *C-C'*. There are two ways to find the *TMD*.

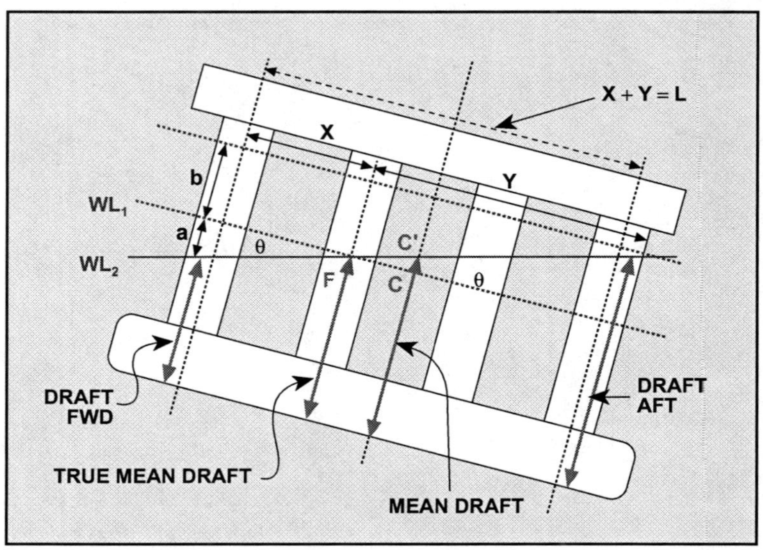

Figure 1.26 Mean draft and true mean draft

$$\text{Total trim (T)} = a + b \text{ and } L = X + Y \text{ (L is the distance between forward and aft draft marks)}$$

$$\text{If } \theta = \text{inclination then } \tan\theta = \frac{T}{L} \text{ but also}$$

$$\tan\theta = \frac{a}{X} \text{ so } \frac{a}{X} = \frac{T}{L}$$

$$a = X\frac{T}{L}$$

$$\text{TMD} = \text{draft forward} \pm \frac{T}{L} \times \text{distance between fwd draft marks and center of flotation}$$

For trim by the head the sign will be minus (–).

Another way is to use the formula: $\text{TMD} = T\dfrac{\text{Distance F to } ⊄}{L}$

1.56 CHANGE OF TRIM (COT) OR CHANGE OF DRAFT (COD) FOR OFF CENTERLINE LONGITUDINAL CENTER OF FLOTATION (LCF)

In case the centerlines of the rig and the change of flotation (COF) do not coincide, the change of trim (COT) on each side will be proportional to the offset of the *longitudinal center of flotation* (*LCF*). This is the same principle as explained for the mean draft and true mean draft. If the rig initial draft were even keel, the COT is Figure 1.27.

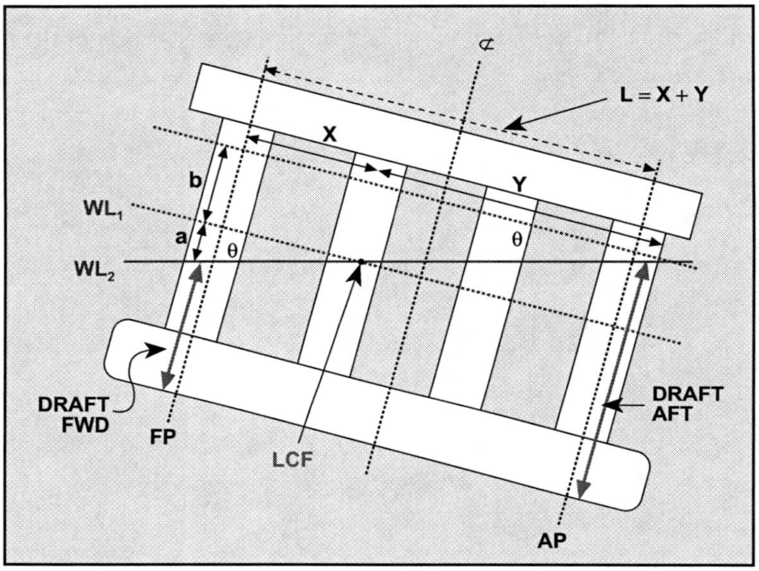

Figure 1.27 COT versus LCF

$$\text{Trim (T)} = a + b \text{ and Length (L)} = X + Y$$

$$\tan\theta = \frac{a+b}{L} \text{ also } \tan\theta = \frac{a}{X} \text{ so}$$

$$\frac{a}{X} = \frac{a+b}{L}$$

$$a = T\frac{X}{L} \text{ and } b = T\frac{Y}{L}$$

$$\text{COT fwd} = T\frac{\text{Distance FP to LCF}}{L}$$

$$\text{COT aft} = T\frac{\text{Distance AP to LCF}}{L}$$

1.57 MOMENT TO HEEL OR TRIM 1° (MH1°-MT1°)—MOMENT TO TRIM OR HEEL 1 CM. (MCT-MCH)

The hydrostatic properties provide the values for the moments to heel or trim the rig 1° or 1 cm. MH1°, MT1°, MCT, and MCH are values at regular draft intervals for a specific value of *VCG*. The example of Figure 1.28 is based on the longitudinal situation.

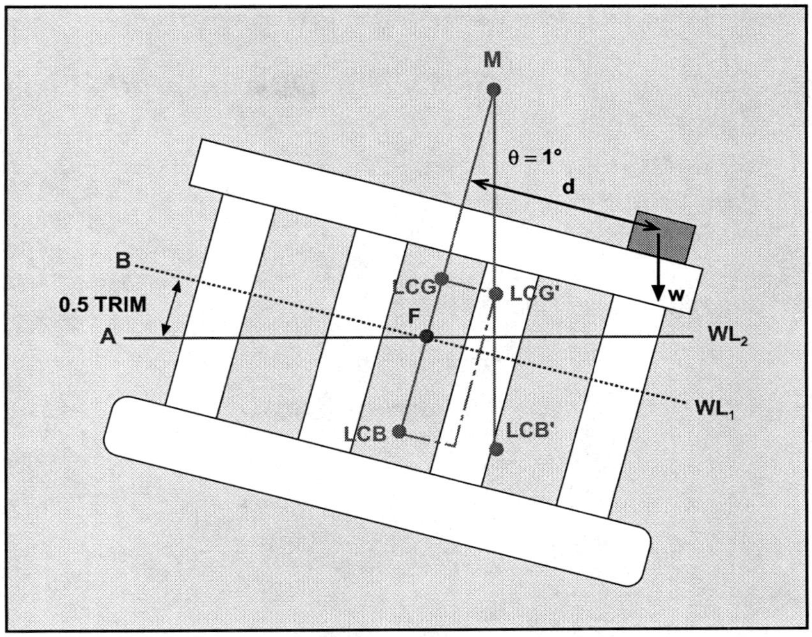

Figure 1.28 Calculation of MT1° and MCT

$$GG' = \frac{w \times d}{W} \quad \text{also } \tan\theta = \frac{GG'}{GM} \rightarrow GG' = GM \tan\theta$$

The value for 1° → $GG' = GM \tan 1°$

Substitute GG' → $\frac{w \times d}{W} = GM \tan 1°$

$$w \times d = W \times GM \tan 1°$$

w × d is specific moment for 1° trim or heel.

$$MT1° = W \times GM \times \tan 1°$$

This is the value in hydrostatic properties based on GM value.

We know $\tan\theta = \frac{COT}{L} \rightarrow COT = L \times \tan\theta$. The trim of 1 cm = $\frac{1}{100}$ m

To trim 1 cm the formula is: $\frac{1}{100}\text{ m} = L \times \tan\theta \rightarrow \tan\theta = \frac{1}{100\,L}$

$$w \times d = W \times GM \tan\theta$$

Substitute tan θ and if w × d is the specific moment to trim 1 cm then:

$$MCT1\text{ cm} = \frac{W \times GM}{100\,L}$$

To change the trim 1" the formula to use is

$$MT1'' = \frac{W \cdot GM_l}{12L}$$

This is the value in hydrostatic properties based on one GM value.

Example. Multiple load/discharge. Find the new drafts.

A drilling rig length is 110 m. The draft is 20 m forward and 21.1 m aft. *TPC* = 6.75 t/cm. *MCT* =75 t•m/cm. The *COF* is 2 m aft of amidships.

130 tonnes is loaded 70 m forward of *COF*.
65 tonnes is loaded 18 m aft of *COF*.
79 tonnes is discharged 31 m forward of *COF*.
Find the new draft and trim.
Find the final mean draft and true mean draft.

First find the moments

Weight	Distance from COF	Moment
+ 130 tonnes	70 m fwd	9,100 t·m fwd
+ 65 tonnes	18 m aft	1,170 t·m aft
– 79 tonnes	31 m fwd	2,449 t·m aft
+ 116 tonnes		5,481 t·m fwd

$$\text{Increase in average draft} = \frac{\text{weight change in tonnes}}{\text{TPC}} = \frac{116}{6.75} = 17 \text{ cm rounded off}$$

$$\text{COT} = \frac{\text{total change of moment}}{\text{MCT}} = \frac{5{,}481 \text{ t·m}}{75 \text{ t·m/cm}} = 73 \text{ cm}$$

$$\text{COD for COT fwd} = \frac{57 \text{ m}}{110 \text{ m}} \times 73 \text{ cm} = 38 \text{ cm}$$

$$\text{COD for COT aft} = \frac{53 \text{ m}}{110 \text{ m}} \times 73 \text{ cm} = 35 \text{ cm}$$

	Fwd	Aft
Initial draft	20.00 m	21.10 m
Increase in draft	+ 0.17 m	+ 0.17 m
	20.17 m	21.27 m
COD for COT	+ 0.38 m	– 0.35 m
Final drafts	20.55 m	20.92 m

$$\text{Trim aft} = 20.92 \text{ m} - 20.55 \text{ m} = 0.37 \text{ cm aft}$$

$$\text{Final mean draft} = \frac{20.55 \text{ m} + 20.92 \text{ m}}{2} = 20.73 \text{ m rounded off}$$

$$\text{True mean draft} = \text{draft fwd} + \frac{T}{L} \times \text{distance between fwd and COF}$$

$$\text{TMD} = 20.55 \text{ m} + \frac{0.37 \text{ m}}{110 \text{ m}} \times 57 \text{ m} = 20.55 + 0.19 = 20.74$$

EXERCISE 3
Chapter 1

1. The initial draft of a drilling vessel is 10' forward and 14' aft. After discharging, the draft is 9' forward and 13' aft. Find the change of average draft and the change of trim (COT).

2. The initial draft of a rig is 12.30 m forward and 12.30 aft. The COF is amidships. Weights are shifted aft to cause a change of trim of .42 m. Find the new draft.

3. A drilling vessel has an initial draft of 19'06" forward and 22'06" aft. After loading and discharging, the final draft is 19'04" and 21'06". The COF is amidships. Find:

   a. The COT
   b. The change of mean draft

4. A drilling vessel is 160 m long. The draft on even keel is 6.35 m. Weights are shifted to change the trim 0.70 m by the stern. The COF is 4.70 m aft of amidships. Find the new drafts forward and aft and the new true mean draft.

5. A semisubmersible with a length of 420 ft and a COF 190 ft forward of the *aft perpendicular* (*AP*) floats at a draft of 16'06" fwd and 15'06" aft. If weights are moved aft to trim the vessel 20" calculate:

   a. The initial trim
   b. The initial mean draft amidships
   c. The initial true mean draft
   d. The final trim
   e. The final draft forward and aft
   f. The COT
   g. The final trim
   h. The final mean draft amidships
   i. The final true mean draft
   j. The change of displacement

6. The length of a semisubmersible is 105 m.The LCF is 1.8 m forward of amidships. The draft is 16.75 m forward and 17.00 m aft. A shift of weight causes a COT of 0.40 m by the bow. Find:

   a. The initial trim
   b. The initial mean draft
   c. The initial true mean draft
   d. The final draft forward and aft
   e. The COT
   f. The final trim
   g. The final mean draft
   h. The final true mean draft

7. a. A vessel with L = 500 ft and LCF = 200 ft forward of the aft perpendicular floats with a draft of 23'08" fwd and 23'04" aft.

   Calculate 1) The initial mean draft
   2) The initial true mean draft

   b. If 208 short tons are moved from a position 180 ft forward of amidships to 148 ft aft of amidships and MT1" is 6,200 ft•ton.

   Calculate 1) The change of trim
   2) The final forward and aft drafts
   3) The final mean draft
   4) The final true mean draft

8. A drilling vessel with L = 114 m and TPC = 17, MCT1 cm = 115. LCF = 1.7 m aft of amidships. The draft is 8.22 m fwd and 8.60 aft. 160 mt is loaded 40 m from the stern. Find:

   a. The initial trim
   b. The initial mean draft
   c. The initial true mean draft
   d. The final draft forward and aft
   e. The COT
   f. The final trim
   g. The final mean draft
   h. The final true mean draft

9. A vessel with L = 400 ft floats at a draft of 16'08" fwd and 18'05" aft. LCF is 220 ft forward of AP. TPI = 59 short tons and MT1" = 1,570 ft•tn. Calculate the final draft forwards and aft if the following cargo has been loaded and discharged.

   Loaded: 161 st at 295 ft forward of AP
   143 st at 185 ft forward of AP
   Discharged: 180 st at 265 ft forward of AP
   124 st at 85 ft forward of AP

10 A drilling vessel with L = 482 ft floats at a draft of 25'02" forward and 25'06" aft. Find the new draft after 176 st is loaded at 124 ft forward of AP. TP1 = 44 st and MT1 = 2,400 ft•tn. LCF is 15 ft aft of amidships.

1.58 CALCULATION OF THE VALUE OF BM

The value of *BM* is calculated over a range of drafts and tabulated in the hydrostatic properties of the rig. Example of *BM* calculation for a barge (fig. 1.29).

$$\text{If inclined B moves over a distance } BB' = \frac{\text{Volume wedge} \cdot \text{Distance between wedge centers } (gg')}{\text{Volume of displacement } (V)}$$

$$\text{Tan } \theta = \frac{X}{0.5b} \rightarrow X = 0.5b \tan \theta \text{ or}$$

$$V \text{ wedge} = L \cdot 0.5b \cdot 0.5 \frac{b \tan \theta}{2} \text{ or } V \text{ wedge} = \frac{L \cdot b^2 \cdot \tan \theta}{8} \quad (1)$$

$$gg' = 2 \cdot \frac{2}{3} \cdot \frac{b}{2} \rightarrow gg' = \frac{2}{3} b \quad (2)$$

Substitute formulas 1 and 2

$$BB' = \frac{(L \cdot b^2 \cdot \tan \theta)\left(\frac{2}{3} b\right)}{8V} \text{ or } BB' = \frac{L \cdot b^3 \cdot \tan \theta}{12V}$$

$$\text{Also } BB' = BM \tan \theta \rightarrow \text{Substitute } BB'$$

$$BM \tan \theta = \frac{L \cdot b^3 \cdot \tan \theta}{12V} \text{ or } BM = \frac{L \cdot b^3}{12V}$$

The second moment about the centerline of the waterplane is called inertia (I).

$$\text{For a rectangular waterplane the moment of inertia } I = \frac{L \cdot b^3}{12}$$

$$\text{So for a rectangular waterplane } BM = \frac{I}{V} = \frac{L \cdot b^3}{12V}. \text{ For a ship-shaped vessel } BM = \frac{\alpha \cdot L \cdot b^3}{12V}$$

$$\text{where } \alpha = \text{a coefficient for a different shape of waterplane}$$

$$\text{For a semisubmersible (fig. 1.29)} \quad BM = \frac{\Sigma \left(\frac{\pi d^4}{64}\right) + Ax^2}{V}$$

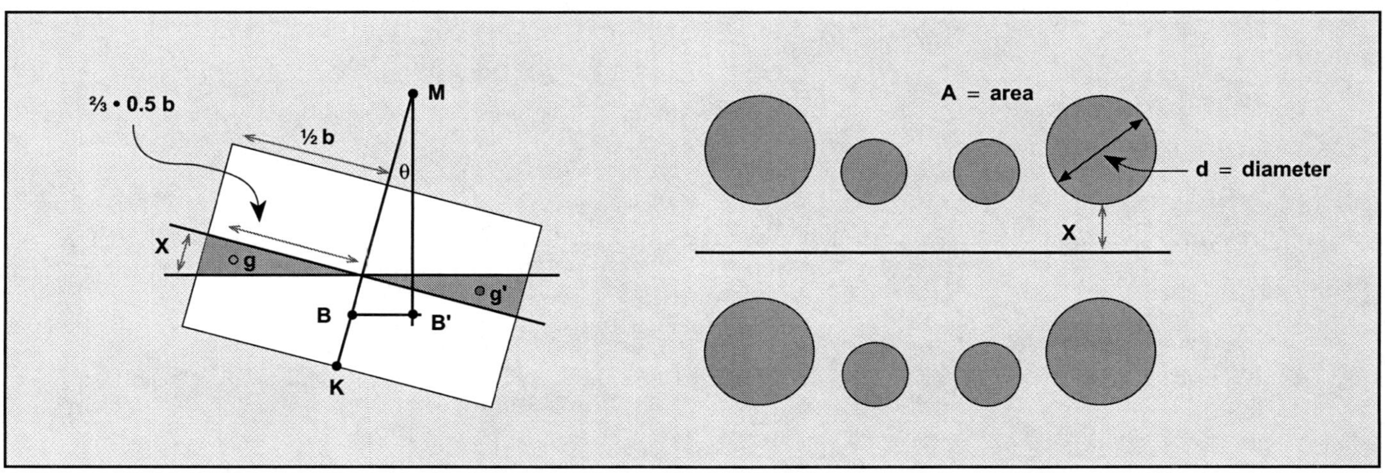

Figure 1.29 BM for barge-shaped vessel—Waterplane semisubmersible deep draft

1.59 THE INCLINING EXPERIMENT

The inclining experiment is carried out to confirm the light weight of the rig with the corresponding actual positions of VCG, LCG, and TCG.

To ensure accurate results, certain conditions are required to perform a good inclining experiment.

1. The displacement of the rig should be as close as possible to the lightship weight.
2. All loose equipment not part of the lightship weight should be discharged.
3. Any loose equipment should be secured and accurately recorded on weight and position.
4. At the start, the rig should be without any heel or trim.
5. All tanks and void spaces need to be inspected.
6. All tanks should be empty and stripped or full. No free surface.
7. The location should be sheltered.
8. The weather should be calm with only light winds and no wave action.
9. The mooring lines should be slack.
10. Except for the inclining crew, all personnel should leave the rig. Any persons to stay on board should remain at their assigned positions.
11. The density of the water in which the rig floats should be measured.

The basic principle of the inclining experiment is to shift a set of known weights several times over a measured distance. To measure the angle, at least two identical plumb lines are installed. Each plumb consists of a long wire with a weight at the bottom. The weight is suspended in a tray with oil to dampen the movements. The deflection is measured at the bottom of the plumb line over a graduated batten. For the purpose of the explanation, only one plumb line is displayed in Figure 1.30. Other modern angle measurement equipment may be used instead of the plumb line. The process of shifting weights is repeated several times to each side of the rig and backwards.

w = weight shift. d = distance of weight shift. W = displacement. θ = angle of plumb line and rig.

$$GG' = \frac{w \cdot d}{W} \text{ (1) and } \tan\theta = \frac{GG'}{GM} \rightarrow GG' = GM\tan\theta \text{ (2). Substitute (2) in (1)} \rightarrow GM\tan\theta = \frac{w \times d}{W} \text{ or}$$

$$GM = \frac{w \cdot d}{W} \cdot \frac{1}{\tan\theta} \quad (3)$$

$$\text{For the plumb line: } \tan\theta = \frac{\text{deflection plumb line}}{\text{length plumb line}} \text{ or } \frac{1}{\tan\theta} = \frac{\text{length plumb line}}{\text{deflection plumb line}} \quad (4)$$

$$\text{Substitute (4) in (3) gives: } GM = \frac{w \cdot d}{W} \cdot \frac{\text{length plumb line}}{\text{deflection plumb line}}$$

Once the value of GM is known, KG can be calculated. The values of KB and BM are known values as calculated by the naval architect. KM = KB + BM → KG = KM − GM.

The result is the VCG or KG for light weight of the rig.

The standards of the IMO Code for the Construction and Equipment of MODUs specifies that:

1. The results of the inclining test should be maintained in the rig's marine operations manual.
2. Records of all changes to machinery and structure that affects the lightship data should be recorded into a logbook.
3. Every five years a deadweight survey should be conducted.
4. If the deadweight survey indicates a change from the calculated lightship weight of more than 1%, an inclining test should be conducted under supervision of a classification society or other recognized authority.

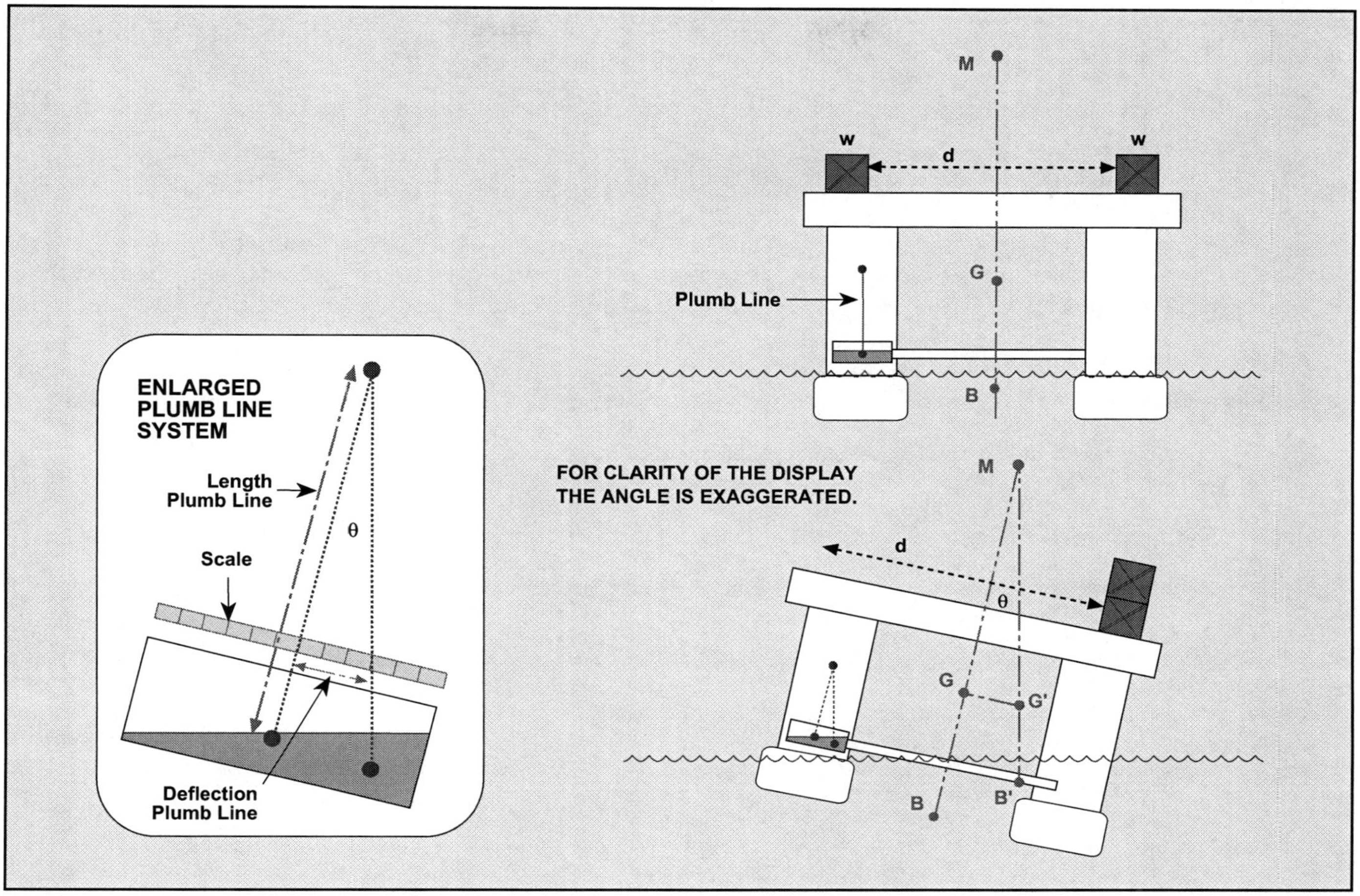

Figure 1.30 The inclining experiment

Below is a typical example of the information contained in the marine operations manual (MOM) for Rig Type 4:

1.6 LIGHTSHIP DATA

The following lightship data is from the April, 1999 inclining experiment at Ingalls Shipbuilding, Pascagoula, Mississippi.

Lightship Weight:	34,135.94 metric tons
Lightship Mean Draft:	4.62 m (15.16 ft)
Lightship VCG:	20.485 m above BL
Lightship LCG:	117.411 m forward of AP
Lightship TCG:	0.201 m port of Centerline

References for the above data are as follows:

VCG:	Hull molded baseline (BL); positive up.
LCG:	Aft Perpendicular (AP) at Frame 0; positive forward.
TCG:	Rig centerline (CL); positive port.

The various classification societies may use more stringent criteria than the IMO Code. For example, for the Rig Type 1 and Rig Type 2 series:

1. The calculated displacement should not exceed 200 st from the displacement corresponding to the draft readings.
2. The calculated TCG and LCG values should be within 1 ft of the actual values. The actual values can be determined by the heel and trim readings.

1.60 THE FREE SURFACE EFFECT

If a tank is completely full (with fluid up into the sounding pipe), the fluid acts as a solid weight. A change of heel or trim will not have any effect on the stability of the rig (see fig 1.31). *B* moves to *B'* but the value of *GM* remains the same.

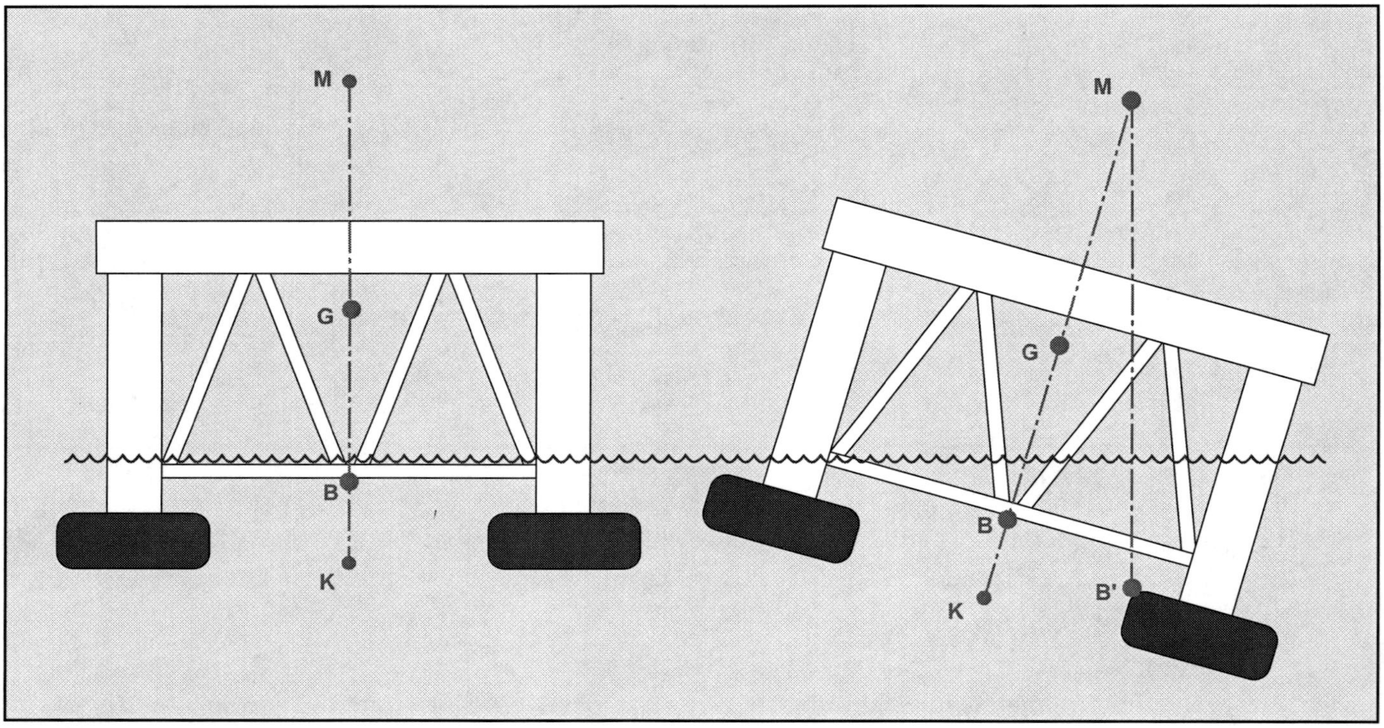

Figure 1.31 Full tanks and heel

The situation changes if a tank is partially filled (slack tank). If the rig moves by external forces, the fluid will move to the low side. It appears as if *G* shifts to *G'* parallel to *g-g'*. The result is that the initial arm of stability *GZ* is now reduced to G_vZ_v. Instead of a righting arm $GZ = GM \sin\theta$, the real value is only $G_vZ_v = G_vM \sin\theta$. The value of *GM* reduced to G_vM (fig. 1.32).

The effect is as if the weight of the water is suspended from a higher point *m* and can swing free like a pendulum.

THE EFFECT OF FREE SURFACE IS A VIRTUAL REDUCTION OF GM

The effect in the longitudinal direction is the same as in the transverse direction. The effect depends on the size of the tank and free surface moment.

The apparent shift of G to G_v can be calculated. The moment of inertia as explained for the waterplane of a vessel has the same effect as the waterplane of the fluid in the slack tanks. The distance BM for the rig's waterplane is now gm for the waterplane in each tank (fig. 1.32). The calculation for GG_v is valid for the metric and Imperial systems.

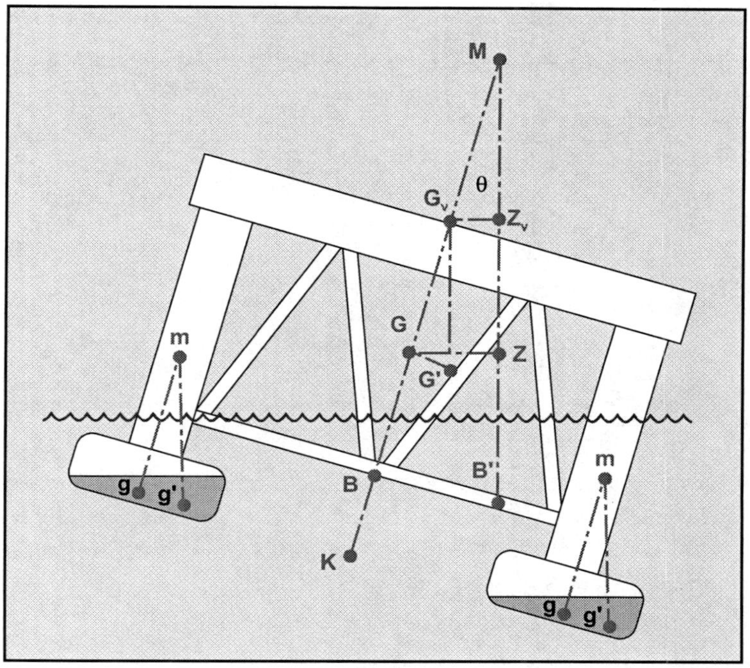

Figure 1.32 Free surface effect

If W = the displacement of the vessel, w = the weight of the fluid, and

d = the distance gm, then:

$$GG_v = \frac{w \times d}{W}$$

d is in fact the distance gm that causes the upward movement of G to G_v.

If v = volume of fluid in the tank, δ_1 = density of fluid in the tank

V = volume of displacement, δ = density of water:

$$gm = \frac{i \text{ tank}}{v}$$

Also W = V × δ and w = v × δ_1 substitute all this:

$$GG_v = \frac{v \times \delta_1 \times i}{V \times \delta \times v}$$

$$GG_v = \frac{i}{V} \times \frac{\delta_1}{\delta}$$

For a rectangular-shaped tank, $i = \dfrac{L \times b^3}{12}$

For transverse free surface effect and a rectangular tank, $GG_v = \dfrac{L \times b^3}{12V} \times \dfrac{\delta_1}{\delta}$

For longitudinal free surface effect and a rectangular tank, $GG_v = \dfrac{b \times L^3}{12V} \times \dfrac{\delta_1}{\delta}$

1.61 THE EFFECT OF BULKHEADS IN TANKS ON THE FREE SURFACE

Because of the third power to the width, the free surface can be reduced by a quarter if a bulkhead through the centerline divides a tank in two. The proof is as follows:

If a tank is divided in two, the free surface is only half the original size.

This means that the width of the tank is now $\frac{1}{2}$ b.

For one tank, $GG_v = \dfrac{L \times \left(\frac{1}{2}b\right)^3}{12\,V}$ or $GG_v = \dfrac{L}{12\,V} \times \dfrac{b^3}{8}$

For two tanks with the same overall width as the initial one tank: $GG_v = 2 \times \dfrac{L}{12\,V} \times \dfrac{b^3}{8}$

$$GG_v = \frac{1}{4} \times \frac{lb^3}{12\,V}$$

Dividing a tank in three reduces the free surface effect by one-ninth and if divided in four the reduction is one-sixteenth. The free surface moment is added to the total moment. Tables 1.3, 1.4, and 1.5 display examples of tank tables with free surface moments for different types of rigs. To distinguish the difference, VCG_{dry} is the value without the free surface effect and VCG_{wet} includes the free surface effect.

1.62 SUMMARY FREE SURFACE EFFECT

1. Any free surface will reduce the stability of the rig.
2. The third power to the width of the fluid is a major factor for the value of GG_v. Width in this respect is the breadth of the tanks for transverse stability and the length of the tank for longitudinal stability.
3. The amount of weight of the fluid is not a deciding factor. The free surface takes effect immediately upon the moment when any water can move freely within the tank. An almost empty tank already has a considerable amount of free surface moment.
4. The density of the fluid affects the free surface effect. Heavyweight mud will produce more free surface effect than salt water.
5. Subdivision by bulkheads in the tanks reduces the free surface effect considerably.
6. Only a tank filled up into the sounding pipe or a stripped tank will have no free surface effect.
7. The location of the tank has no influence on the free surface effect.
8. The total free surface effect is the sum of the free surface effects of each tank.

Example. Find the VCG including free surface correction.

The stability calculation final result is:

VCG dry = 47 ft. Displacement 18,950 st. Total free surface effect 9,045 ft•st. What is the VCG wet value?

W	VCG		Moment
18,950 st	47 ft		890,650 ft•t
		Free surface effect	9,045 ft•st
18,950 st	?		899,695-ft•st

$$\text{VCG wet} = \frac{899{,}695 \text{ ft}\cdot\text{st}}{18{,}950 \text{ st}}$$

VCG wet = 47.5 ft

The free surface effect is rise in G to G_v = 0.5 ft

TABLE 1.3

Example Rig Type 2A Free Surface Effect of Tanks

RIG TYPE 2A

FREE SURFACE EFFECT OF HULL TANK LIQUIDS

Sounding (FT)	PB1A SB10A	SB1A PB10A	PB1B SB10B	SB1B PB10B	B2,B3 B6,B9	B4,B5 B7,B8	DW1 DW2	FO
			FSE (ft-ST) about Longitudinal Axis (Roll)					
0	0	0	0	0	0	0	0	0
1	290	777	227	227	970	847	826	818
3	499	1098	421	421	1428	1247	1216	1204
6	690	1380	596	596	1850	1615	1576	1560
9	776	1505	677	677	2050	1791	1747	1729
10.5	783	1515	682	1454	2074	1812	1767	1749
12	784	784	683	1456	2050	1791	1747	1729
15	744	744	646	1404	1850	1615	1576	1560
18	633	633	543	1251	1428	1247	1216	1204
20	477	477	398	1037	970	847	826	818
21	0	0	0	0	0	0	0	0
			FSE (ft-ST) about Transverse Axis (Pitch)					
0	0	0	0	0	0	0	0	0
1	974	1615	630	630	3522	2355	2297	2971
3	1443	2011	1144	1144	4013	2684	2618	3385
6	1630	2195	1336	1336	4378	2928	2856	3693
9	1672	2245	1345	1345	4533	3032	2957	3823
10.5	1681	2254	1354	2025	4550	3043	2968	3837
12	1678	1678	1384	2022	4533	3032	2957	3823
15	1666	1666	1375	2002	4378	2928	2856	3693
18	1605	1605	1321	1922	4013	2684	2618	3385
20	1526	1526	1258	1809	3522	2355	2297	2971
21	0	0	0	0	0	0	0	0
		FSE (ft-ST) about Diagonal Axis 49 Deg from Longitudinal						
0	0	0	0	0	0	0	0	0
1	680	1254	457	457	2424	1706	1664	2044
3	1037	1618	833	833	2900	2066	2015	2446
6	1225	1844	1017	1017	3290	2363	2305	2775
9	1286	1926	1057	1057	3464	2498	2436	2922
10.5	1294	1936	1065	1779	3484	2513	2451	2938
12	1293	1293	1082	1778	3464	2498	2436	2922
15	1269	1269	1061	1745	3290	2363	2305	2775
18	1187	1187	986	1633	2900	2066	2015	2446
20	1074	1074	888	1477	2424	1706	1664	2044
21	0	0	0	0	0	0	0	0

Notes:

1. To obtain the reduction in allowable VCG due to excess free surfaces, enter the tabulated values for each tank containing liquid on Form 3. Each value is the free surface effect for one tank.
2. For moderate weather conditions use roll FSE for the draft range 18 to 72 ft and use diagonal axis FSE for the draft range 72 tc 95 ft.
3. For storm conditions use roll FSE for the draft range 55 to 61 ft and use diagonal axis FSE for the draft range above 61 ft.

TABLE 1.4

Example Rig Type 4 Free Surface Effect of a Double Bottom Tank

CUSTOMER - SONAT O. DR. INC
DESCRIPTION OF SHIP - D. P. DRILLING UNIT

CAPACITY TABLES
===============

DENSITY OF THE LOAD (DN) = 1.025 T/M3
STEEL DISCOUNT FACTOR = 2.000 %

COMPARTMENT 26' W.B. DOUBLE BOTTOM TANK, FR.4-33 C.

HEIGHT	AREA FLOT.	VOLUME NET	WEIGHT	C. OF GRAVITY VCG	LCG	VOLUME TCG	I.T.*DN
M	M2	M3	T	M	A.P. M	M	M*T
0.250	165.13	37.2	38.2	0.13	21.24	-0.83	10046.4
0.500	208.22	85.8	87.9	0.27	20.49	-0.72	12536.2
0.750	247.24	142.6	146.2	0.41	19.92	-0.65	14759.9
1.000	285.50	207.3	212.5	0.56	19.42	-0.60	16900.0
1.250	322.95	280.1	287.1	0.71	18.93	-0.55	18947.4
1.500	355.24	360.2	369.2	0.86	18.49	-0.51	20596.7
1.750	349.19	446.9	458.1	1.01	18.08	-0.48	20579.7
2.000	374.58	538.6	552.1	1.15	17.71	-0.46	21706.8
2.250	401.13	634.8	650.6	1.30	17.38	-0.44	22801.8
2.500	413.62	734.8	753.1	1.45	17.09	-0.42	23589.1
2.750	417.50	836.6	857.5	1.59	16.86	-0.41	24209.4
2.900	419.77	898.1	920.6	1.68	16.74	-0.40	24578.0
2.979	420.95	930.7	953.9	1.72	16.68	-0.40	24771.0
2.980	229.39	931.1	954.3	1.72	16.68	-0.40	9084.9
3.000	229.53	935.6	959.0	1.73	16.65	-0.39	9105.4
3.250	231.29	992.0	1016.8	1.81	16.28	-0.37	9364.7
3.500	233.05	1048.9	1075.1	1.89	15.95	-0.35	9629.0
3.750	234.00	1106.0	1133.7	1.98	15.65	-0.33	9897.4
4.000	0.00	1163.6	1192.7	2.07	15.38	-0.32	0.0

TABLE 1.5

Example Rig Type 1 Free Surface Effect of Hull Tanks

For Longitudinal Conditions

FREE SURFACE EFFECT OF HULL TANK LIQUIDS

(CORRECTIONS FOR GM_L)

Percentage of Tank Full Tank \ Moment	20% (t-m)	25% (t-m)	30% (t-m)	35% (t-m)	40% (t-m)	50% (t-m)	55% (t-m)	60% (t-m)	65% (t-m)	75% (t-m)	85% (t-m)	95% (t-m)
Fuel Oil 1	734	211	211	211	211	211	211	211	129	129	129	108
Fuel Oil 2	734	211	211	211	211	211	211	211	129	129	129	108
Day Tank 1	1	1	1	1	1	1	1	1	1	1	1	1
Day Tank 2	1	1	1	1	1	1	1	1	1	1	1	1
Dirty Oil Tank	1	1	1	1	1	1	1	1	1	1	1	1
Drill Water 1	857	247	247	247	247	247	247	247	151	151	151	126
Drill Water 2	857	247	247	247	247	247	247	247	151	151	151	126
Brine Tank 1	2	2	2	2	2	2	2	2	2	2	2	1
Brine Tank 2	3	3	3	3	3	3	3	3	3	3	3	2
Brine Tank 3	2	2	2	2	2	2	2	2	2	2	2	2
Base Oil	3	3	3	3	3	3	3	3	3	3	3	3
Potable Water	382	381	1,091	1,091	1,088	1,083	1,099	1,092	1,084	1,029	1,029	1,029
Void Space Aft	112	105	105	105	103	103	103	103	103	85	93	35
Void Space Fwd	101	70	70	70	68	68	68	68	68	68	73	73
Void Space High Aft	1	2	2	3	3	3	3	3	3	3	3	3
RMT 1	72	72	72	72	72	72	72	72	72	72	72	72
RMT 2	74	74	74	74	74	74	74	74	74	74	74	74
RMT 3	74	74	74	74	74	74	74	74	74	74	74	74
RMT 4	72	72	72	72	72	72	72	72	72	72	72	72
Mud Tank 1	111	111	111	111	111	111	111	111	111	111	111	111
Mud Tank 2	111	111	111	111	111	111	111	111	111	111	111	111
Mud Tank 3	111	111	111	111	111	111	111	111	111	111	111	111
Mud Tank 4	111	111	111	111	111	111	111	111	111	111	111	111
Mud Tank 5	7	7	7	7	7	7	7	7	7	7	7	7
Mud Tank 6	459	459	459	459	459	459	459	459	459	459	459	459
Mud Tank 7	7	7	7	7	7	7	7	7	7	7	7	7
Mud Tank 8	459	459	459	459	459	459	459	459	459	459	459	459
Mud tank 9	111	111	111	111	111	111	111	111	111	111	111	111
Mud Tank 10	111	111	111	111	111	111	111	111	111	111	111	111
Mud Tank 11	110	110	110	110	110	110	110	110	110	110	110	110
Mud Tank 12	110	110	110	110	110	110	110	110	110	110	110	110

EXERCISE 4
Chapter 1

1. Find the angle caused by a weight shift on a barge if GM = 6.0 ft and G' after the weight shift is 5 inches of the centerline.

2. The lightship weight of a vessel is 6,360 st. A weight of 12 st is moved 49 ft from amidships to port. The deflection of a 40 ft pendulum is 14". KM = 26.88 ft. Calculate the GM and KG value for lightship.

3. A drilling vessel is ready for the inclining experiment. The displacement during the inclining test includes the following additional weights not part of the lightship weight of 34,120 mt.

   a. 750 mt of ballast in full tanks with a KG = 2.60 m
   b. 950 mt loose equipment and cargo with a calculated KG = 26.8 m
   c. Total inclining weights 120 mt with KG = 24.9 m

   Inclining weights = 32 mt are moved over 36 m. The plumb line with a length of 7.9 m shows a deflection of 8.9 cm. Calculate the final VCG and GM_t for lightship condition if during the inclining test KM_T = 27 m, and after removing the additional weights, equipment, and cargo, KM_T = 27.8 m.

4. What is the displacement of a vessel if a weight of 12 mt is shifted 10.8 m from port to starboard? The deflection of a plumb line of 8.20 m is 20 cm. GM = 0.75 m.

5. A vessel of 8,190 st is floating in seawater. If a tank 60 ft long and 54 ft wide is partly filled with seawater, calculate the reduction of GM with respect to roll due to the free surface effect.

   a. For the tank without division
   b. For the tank with a longitudinal centerline bulkhead.

6. Use Table 1.3 Rig Type 2A. The tank soundings are PB1A = 9 ft, SB1B = 2 ft, DW1 = 19 ft, DW2 = 20.5 ft, SB10B = 1 ft, and PB10B = 4 ft. The VCG dry = 61.4 ft. Displacement is 25,700 st, KB = 20 ft, KM = 67 ft. Calculate the GM wet with diagonal values.

7. The displacement of a rig is 18,500 mt. GM = 3.4 m. A total of 420 mt is shifted from port to starboard over a distance of 15 m. Calculate the angle of heel without free surface. What will the heel be if two tanks each with L = 12 m and width = 14 m have a free surface?

1.63 THE EFFECT OF WEIGHT SHIFT ON THE CENTER OF GRAVITY

If a *weight* (W) is moved over a distance d, the center of gravity will move from G to G' parallel to the movement of d (fig. 1.33). Because of the moment $w \cdot d$, the rig will heel in this case to the right. The rig stabilizes under an angle with a new position of G'.

$$W \cdot GG' = w \cdot d \text{ or}$$

$$GG' = \frac{w \cdot d}{W}$$

W is the weight of the rig

d is the distance of the weight move

w is the weight of the load

The heel or trim moment is

$w \cdot d$ or $W \cdot GG'$

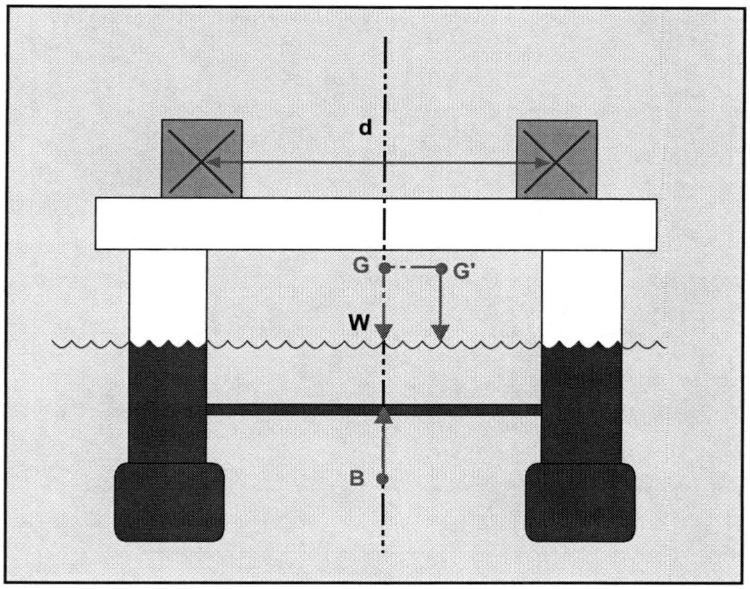

Figure 1.33 Shift of G to G'

After the weight move, the rig stabilizes with an angle θ. The points B, G', and M are on a vertical line (fig 1.34). The opposite forces through B and G' are in line; i.e., there is no couple. GM is positive. The angle θ is calculated as shown to the right.

The angle θ is:

$$\tan \theta = \frac{GG'}{GM}$$

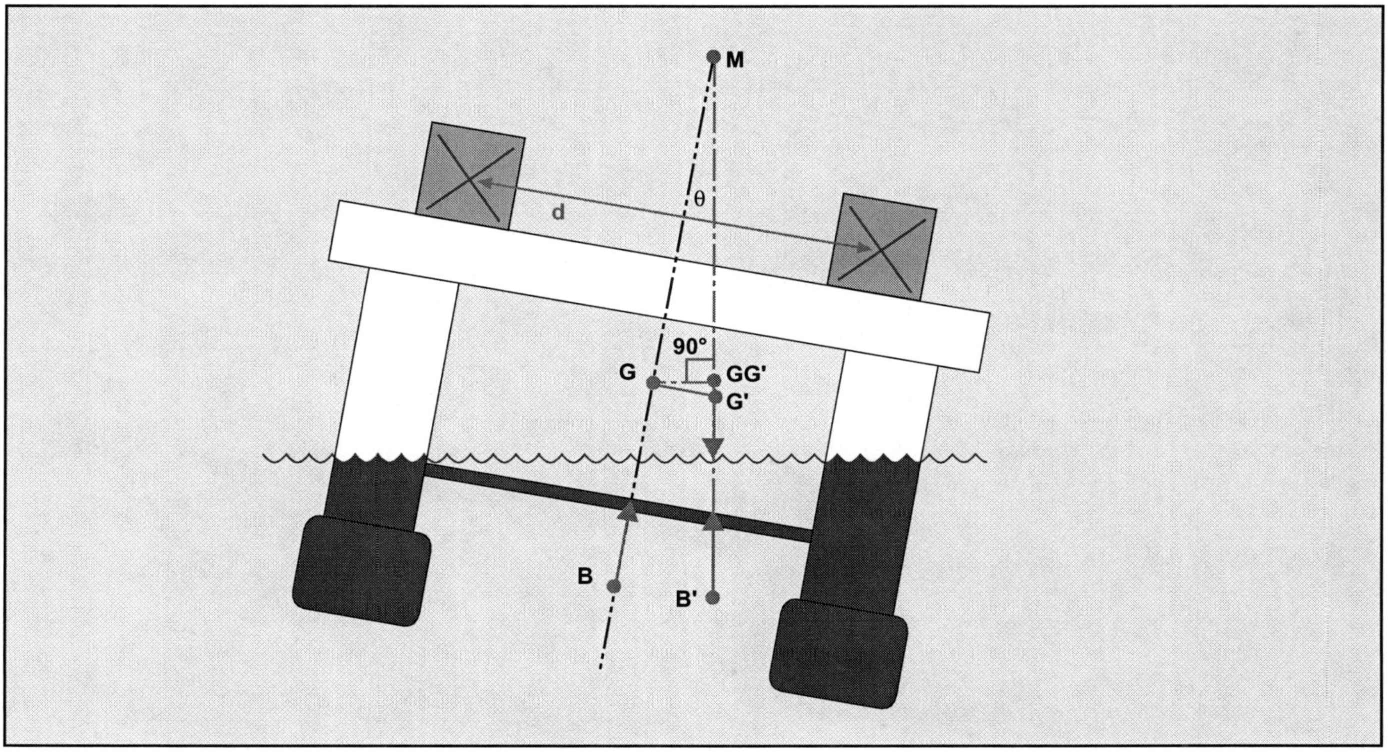

Figure 1.34 Shift of B and G

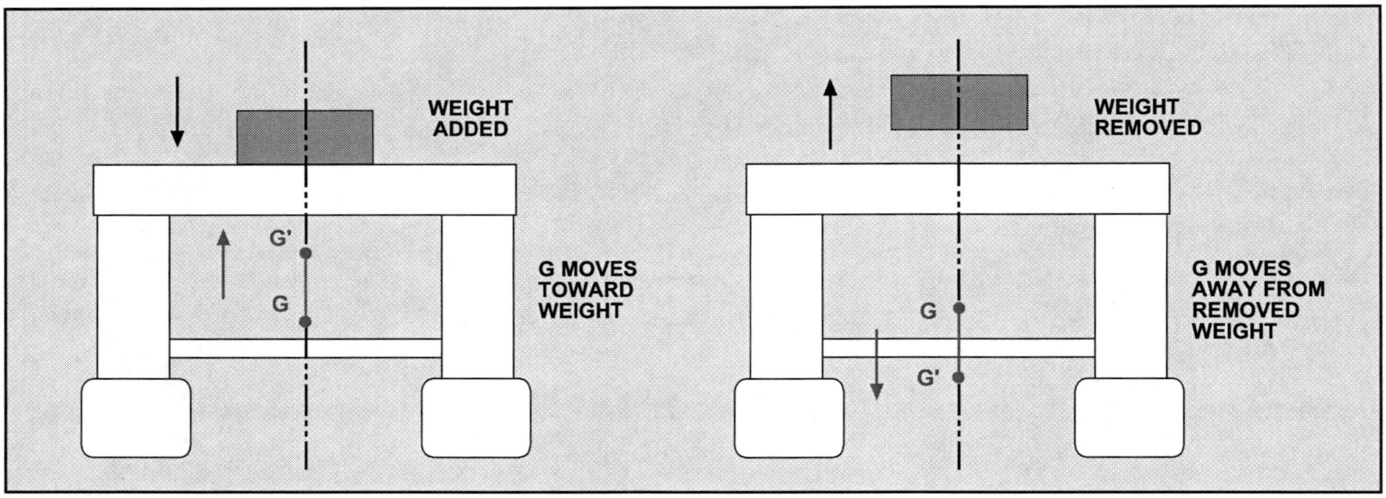

Figure 1.35 **Shift of G to G'**

A weight change will affect the position of *G* in various ways:

1. Discharging a weight results in a new position of *G* directly away from the removed weight (fig 1.35).
2. Loading a weight results in a new position of *G* directly towards the added weight (fig. 1.35).
3. Shifting a weight results in a new position of *G* parallel to the weight shift .

Example. Multiple weight changes

A rig of 12,000 st with a *KG* of 18 ft loads 150 st 14 ft above keel, discharges 1,200 st 10 ft above keel and 150 st is shifted from 10 ft above keel to 16.8 ft above keel.

Item	Weight	KG (VCG)	Vertical Moment
Rig	12,000 st	18 ft	216,000 st•ft
Load	+150 st	14 ft	+2,100 st•ft
Discharge	-1,200 st	10 ft	-12,000 st•ft
Shift	±150 st	6.8 ft	+1,020 st•ft
Totals	10,950 st	New *KG*	207,120 st•ft
The new KG value = 207,120 st•ft ÷ 10,950 st = 18.91 ft			

It is more convenient to use the calculations with tabulated moments instead of the single weight shift formula. The daily stability calculations on the rig are done with the tabulated moments as shown above.

Example. Lift of weight. Transverse movement G-G"

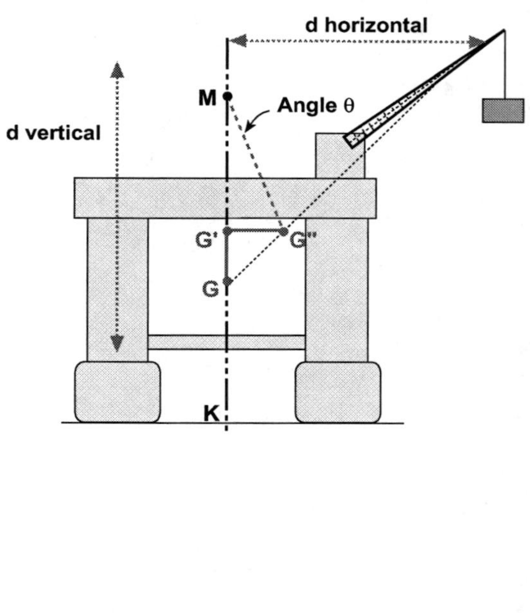

The height of the boom is 60 m above *K*. The weight of the lift is 35 t. The horizontal distance between the centerline and the lift is 40 m.The displacement is 16,500 t and *KG* is 14.50 m. Initial *KM* = 17 m. Calculate the shift of *G* to *G"* and the inclination.

$$\text{Vertical move } GC' = \frac{w \times d}{W + w} \rightarrow GC' = \frac{35\text{ t} \times (60\text{ m} - 14.5\text{ m})}{16{,}500\text{ t} + 35\text{ t}}$$

$$GC' = 0.10\text{ m}$$

$$\text{New KG} = \text{initial KG} + GC' \rightarrow \text{new KG} = 14.5\text{ m} + 0.10\text{ m} = 14.60\text{ m}$$

$$G'M\text{ final} = KM - KG \rightarrow G'M\text{ final} = 17\text{ m} - 14.60\text{ m} = 2.40\text{ m}$$

$$\text{Horizontal move} = \frac{35 \times 40}{16{,}500 + 35}$$

$$G'G'' = 0.09\text{ m}$$

$$\tan\theta = \frac{G'G''}{G'M} = \frac{0.09\text{ m}}{2.40\text{ m}} = 0.0375$$

$$\theta = 2.147\text{ degrees}$$

1.64 SUMMARY OF STATIC STABILITY FOR SMALL ANGLES (GENERALLY LESS THAN 8°)

1. During the design and building period, the naval architect calculates the values of KB and BM for a range of drafts.
2. The value of KM can be calculated because KM = KB + BM.
3. The values for KM are tabulated for KM_L longitudinal and KM_T transverse.
4. G is the final position of the center of gravity of the rig. The actual KG or VCG is calculated on board.
5. The starting point is KG lightship and moment.
6. Additional loads and load changes are calculated by subtracting or adding the weights and corresponding moments. The final KG is the total of all moments divided by the total of all weights.
7. GM is used as an indicator of the amount of initial stability.

GM = KB + BM - KG.

8. The hydrostatic properties are tabulated information for the draft range between lightship and loaded condition.
   a. The draft and corresponding displacement
   b. The positions of M, B, center of flotation, all in the longitudinal and transverse planes
   c. The waterplane areas
   d. KM_L and KM_T.
   e. Ton per cm or inch
   f. The moments to change the trim one cm or one inch.

EXERCISE 5
Chapter 1

1. The draft of a box-shaped barge is 19 ft and KM = 22.5 ft. What is the value of BM?

2. A crane with a boom height of 60 m above keel and a reach of 42 m to starboard from the centerline lifts a weight of 24 mt.The displacement is 18,440 mt. KG = 15.80 m. Calculate the shift of G. Find the angle to starboard if KM = 21 m.

3. Find the height of KM transverse for a box-shaped barge with a beam of 20 m, a length of 130 m, and a draft of 6.8 m.

4. A rig floats with a displacement of 13,225 mt. KG = 12.34 m. To prepare the drilling operations, the following cargo is loaded and discharged:

   Loaded:
   a. 350 mt with KG = 14.5 m
   b. 1,250 mt with KG = 5 m
   c. 30 mt with KG = 25 m
   d. 950 mt with KG = 15.5 m
   e. 150 mt with KG = 12.1m

   Discharged:
   a. 560 mt with KG = 13.25 m
   b. 330 mt with KG = 6.01 m
   c. 520 mt with KG = 16.90 m
   d. 150 mt with KG = 11.12 m

   Find the new KG.

5. A barge 75 m long and 12 m wide floats at a draft of 3.5 m. Calculate the increase of KM transverse if the beam of the barge is increased by 1.8 m.

6. A vessel of 4,800 st displacement, KG = 12.6 ft. KM = 16.4 ft. The list is 4° to starboard. Find the angle of list after the following has been loaded and discharged.

   Loaded:
   a. 80 st at 16.5 ft above keel and 12 ft to port of centerline
   b. 40 st at 20.4 ft above keel and 26 ft to starboard of centerline

   Discharged:
   a. 60 st at 4.2 ft above keel and 8 ft to port of centerline
   b. 60 st at 2.4 ft above keel and 15 ft to starboard of centerline

EXERCISE 6
Chapter 1

1. Define the center of buoyancy.

2. In Table 1.2, Rig Type 1 Hydrostatic Tables TCB is 0.00 m. What does this tell you?

3. Explain the difference between the height of the metacenter and the metacentric height.

4. Give the description of the metacenter.

5. Explain why GM is a factor to indicate the amount of stability.

6. Explain why KM_T and KM_L normally do not have the same value.

7. Under what condition will KM_T and KM_L, have the same value?

8. Define the stability couple.

9. Which factors determine the amount of the righting moment?

10. The center of mass of a drilling vessel is also known as the:
    a. center of buoyancy.
    b. center of gravity.
    c. metacenter.
    d. center of flotation.

11. Define the center of flotation.

12. The position of the center of flotation depends on?

13. What is the stability condition for a KG = 20 ft, KB = 5 ft, and BM 14 ft?

14. Explain the difference between a stiff and tender rig.

15. What will happen to G and B when a vessel is inclined due to some external forces?

16. Explain the conditions:
    a. stable equilibrium
    b. neutral equilibrium
    c. unstable equilibrium

CHAPTER 2
Environmental Forces

2.1 INTRODUCTION

The objective of this chapter is to introduce the fundamental concepts of the effect on stability caused by environmental forces on a floating MODU. The environmental data is an important part of the rig design criteria to meet the required stability criteria.

2.2 ENVIRONMETAL FORCES

The environmental forces acting on a MODU are caused by:

1. waves.
2. wind.
3. current.

2.3 MOTIONS

The floating drilling unit will respond on a single or a combination of the environmental forces with six types of motions (fig. 2.1):

Three rotational motions:
1. Roll
2. Pitch
3. Yaw

Three translational motions:
1. Surge
2. Sway
3. Heave

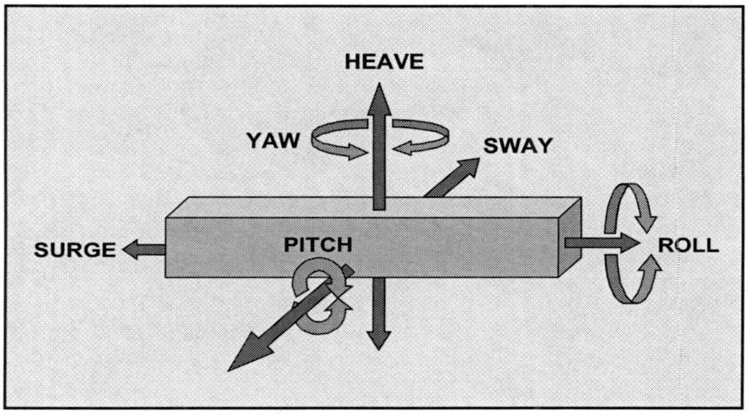

Figure 2.1 Rig motions

2.4 WAVE FORCES

All waves other than waves caused by earthquakes are generated by wind. The wind changes the water surface from ripples of 0.1 m (.25 ft) for light winds to exceptionally high waves of more than 14 m (46 ft). There are two types of waves:

1. Wind waves are waves generated by the wind at the location.
2. Swells are waves away from the area they were generated.

2.5 WAVE PROFILES

The explanation of the wave theory is based on a regular wave profile (fig. 2.2).

A regular wave is defined by

1. The *wave height* (*H*) – The vertical distance between the top of a crest and the bottom of the trough in metres or ft. It is twice the *amplitude* which is the distance from still waterline to the top of the crest or the bottom of the trough.

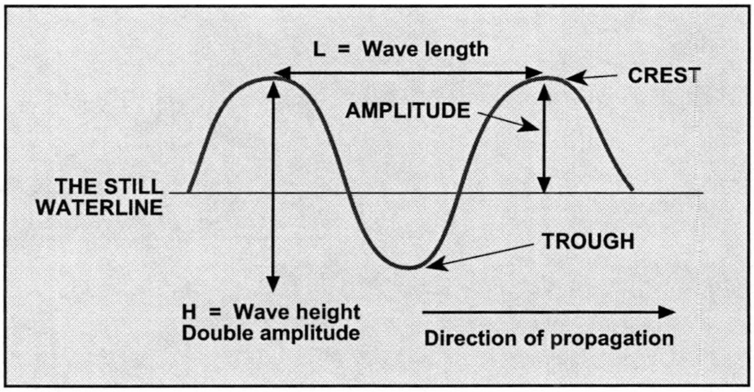

Figure 2.2 Wave profile

2. The *wave period* (T) – The time between two successive crests in seconds.
3. The *wave length* (L) – The horizontal distance between two successive crests in seconds.
4. The *wave velocity* (V) – The velocity of the wave crest in m/sec or ft/sec.
5. *Wave steepness* (H/L) – The ratio between wave height and length.

2.6 WAVE FORMULA

FREQUENCY

$$f = \frac{1}{T}$$

f = wave frequency

T = wave period between two crests

ANGULAR FREQUENCY

$$\omega = \frac{2\pi}{T}$$

WAVE CREST VELOCITY

$$V_{max} = \frac{L}{T} \text{ or } V_{max} = \frac{H}{2}\,\omega$$

V = the maximum velocity of the wave crest

L = wave length

T = wave period

WATER PARTICLE ACCELERATION

$$\alpha = \frac{H \cdot \omega^2}{2}$$

α = maximum acceleration in still water

H = wave height (double amplituded)

ω = angular wave frequency $\frac{2\pi}{T}$

A 60 knot gale can generate waves with a period of 17 seconds and a wave length of 450 m which produces a wave speed of 97 km/hr.

Examples.

What is the wave velocity of a wave with a length of 150 ft and a wave period of 12 sec?

$$V = \frac{150 \text{ ft}}{12 \text{ sec}} \rightarrow V = 12.5 \text{ ft/sec}$$

What is the wave crest velocity of a 30 ft wave with a period of 11 seconds?

$$V = \frac{H}{2} \cdot \omega = \frac{30 \text{ ft}}{2} \cdot \frac{2\pi}{11 \text{ sec}} = \rightarrow V = 8.56 \text{ ft/sec}$$

2.7 WAVE LOADS

Although the wave form moves, the water particles making up the wave only oscillate. The sequence of the circular motions of the individual water particles along the wave length produces the image of the progressing wave. The wave is a standing wave. The orbit of the water particles changes from circular to elliptical from deep water to shallow water. In deep water the circular orbit decays with the water depth (fig. 2.3).

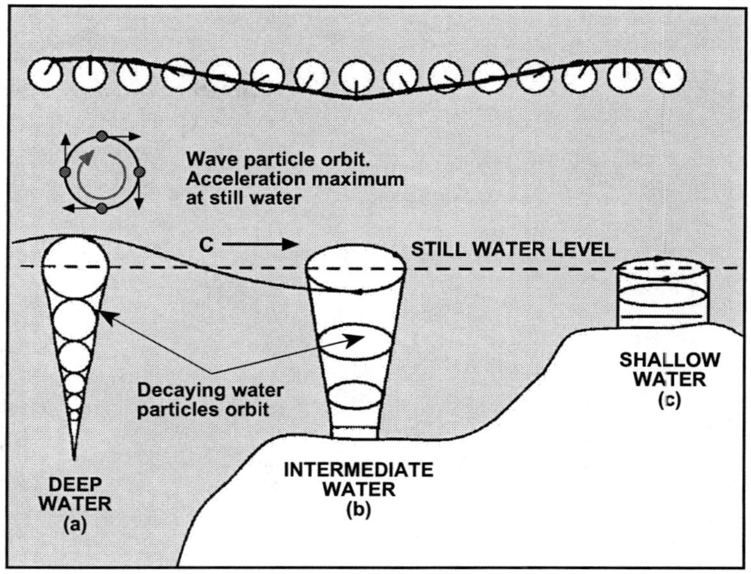

Figure 2.3 Wave particle movement in deep and shallow water

The decaying orbit of the water particles is zero at a depth of half the wave length. Long waves (swell) extend, therefore much deeper than short waves (wind waves). A swell wave with a length of 100 m will still stir the seabed at 50 metres.

The wave action and related forces on floating and fixed structures are complicated to calculate. Different theories are required for deep and shallow waters. The wave shape, height, and direction are other important factors. Wave forces acting on a submerged structure are composed of two major components:

1. The inertia force (F_i).
2. The drag force (F_d).

The classification societies use the various wave theories as applicable for jackups and semisubmersibles for shallow and deep water. The Morison equation is an example to compute wave forces with the following formula:

The Morison equation uses two forces to calculate the wave forces

1. The inertia force (F_i) → $F_i = C_i \times \rho \times V \times a$. F_i is maximum at 90° and 270°.

2. The drag force (F_d) → $F_d = 0.5\ C_d \times \rho \times A \times v^2$. F_d is maximum at 0° and 180°.

C_i = coefficient of inertia (1.8 for a cylinder)

ρ = density of seawater

V = volume of the body

a = acceleration of the water particles around the body

C_d = drag coefficient (ranges from 0.5 to 0.9 for a cylinder)

A = projecting area of the body

v = velocity of the water particles around the body

The water particle acceleration is maximum at zero crossing (still water level) at 90° and 270° and minimum at crest and trough at 0° and 180°. F_i is maximum at still water level. The inertia force is 90 degrees out of phase with the drag force. This means the horizontal speed is maximum at the top and the bottom of the wave at 0° and 180°. F_d is maximum at the top or crest of the wave. Marine growth will cause higher drag forces because of the increase in diameter and drag force coefficient. On some rigs, a graph in the rig's marine operations manual (MOM) displays the results of the wave drift force calculations (fig. 2.4).

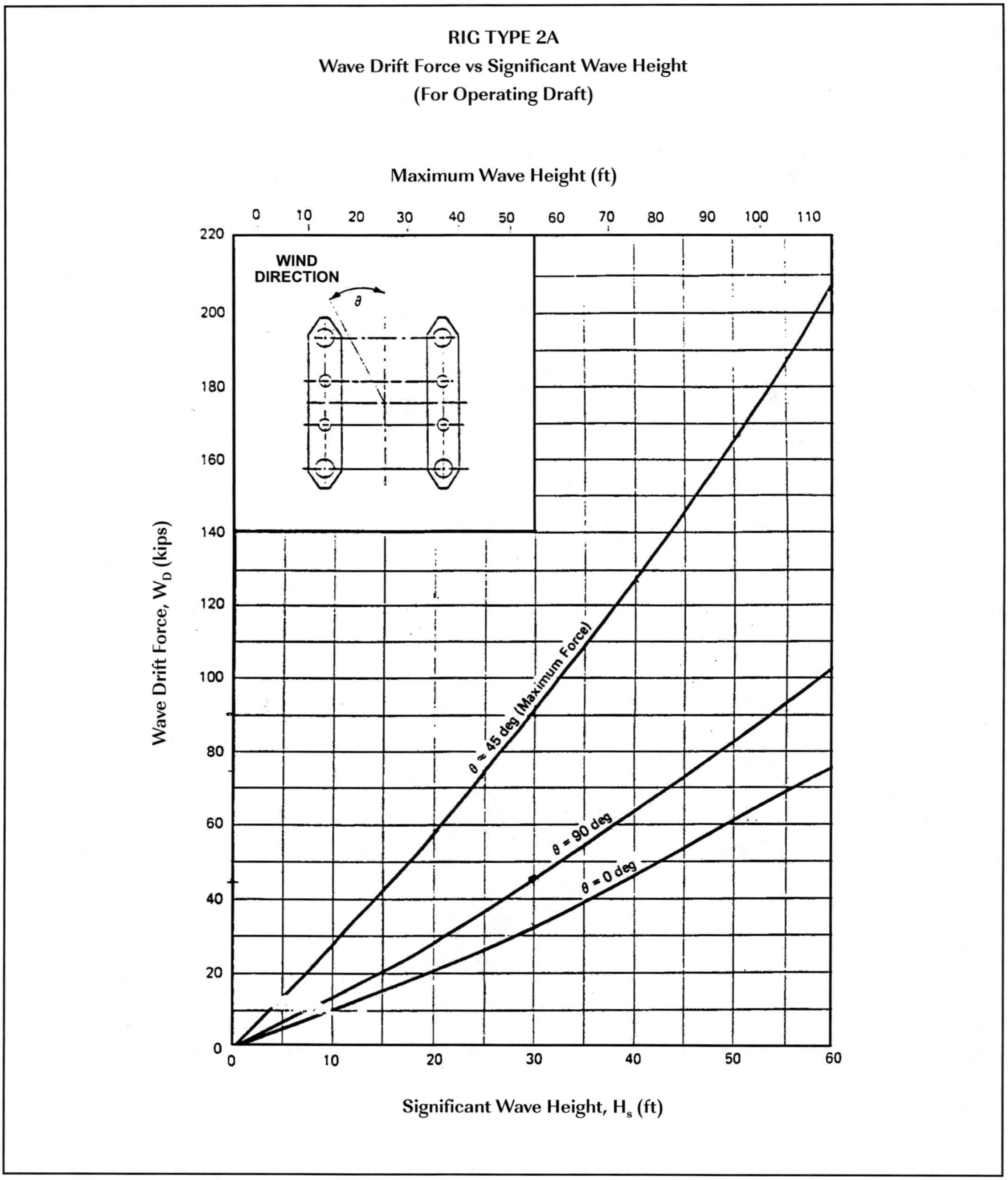

Figure 2.4 Wave forces versus significant wave height for Rig Type 2A

2.8 DESCRIPTION OF SEA STATE

Unfortunately, the waves at sea are not regular waves but a combination of many waves.

The sea state is a combination of many waves. Wind waves and swell spread over a range of frequencies and speeds. The summation of all the waves is called the spectral sea state (fig. 2.5).

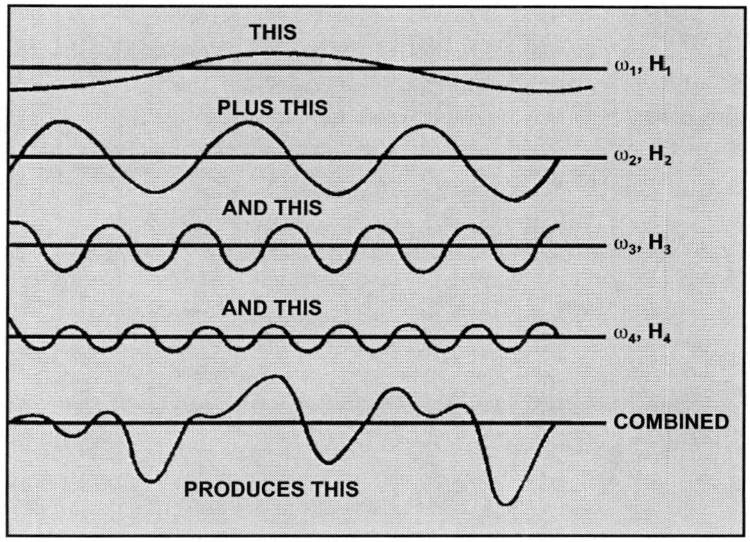

Figure 2.5 Wave spectrum—summation of all waves

2.9 SIGNIFICANT WAVE HEIGHT (Hs) AND MAXIMUM WAVE HEIGHT (Hm)

To describe the spectral sea, the significant wave height is used. The *significant wave height* (*Hs*) is the average height of the highest one-third of the waves in the wave spectrum.

The significant wave height is used to calculate the environmental criteria for a rig. The *maximum wave height* (*Hm*) can be computed from the significant wave height by:

$$Hm = 1.86\ Hs$$

Figure 2.5 is an example of a graphic display of the wave drift force versus the significant wave height of the Rig Type 2A at operations draft. These types of graphs are sometimes part of the rig's marine operations manual.

2.10 WIND DRIVEN WAVES VERSUS SWELL DRIVEN WAVES

The height of the wind wave depends on three factors:

1. The strength of the wind.
2. The time the wind blows from the same direction.
3. The fetch.

The wind fetch is the distance the wind travels uninterrupted with constant speed and direction. The rig in figure 2.6 is protected by short fetches except winds from the NE direction. Only winds from the NE direction can develop to the maximum height.

The height of the wind wave decreases very fast after the wind stops blowing, but the swell continues to travel over a long distance if not disturbed by any landfall.

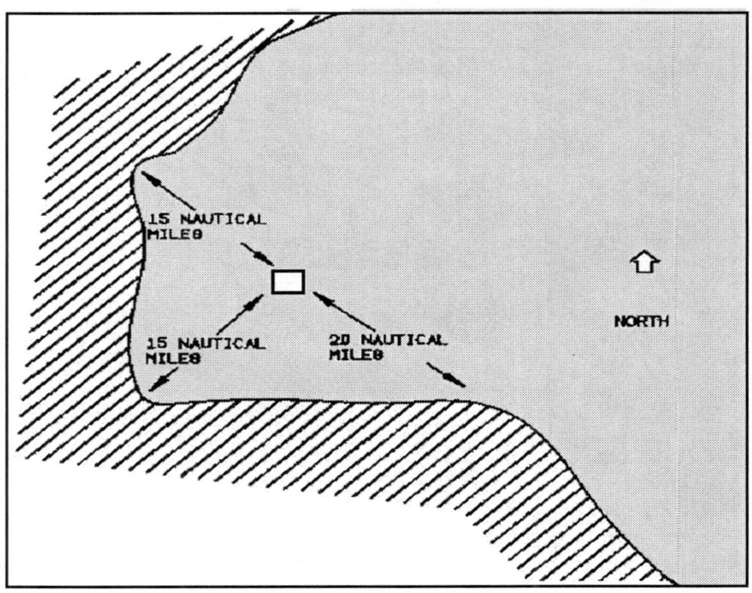

Figure 2.6 Fetch distances in sheltered area

The period of a swell driven wave is always larger than the period of a wind driven wave of the same height. The graph in Figure 2.7 shows the difference in period, between a wind wave and a swell wave of a significant wave of 30 ft, is 12 sec versus 15.5 sec.

2.11 THE BEAUFORT SCALE FOR WIND FORCE AND WAVE HEIGHT

The Beaufort wind scale, developed in 1805, is still used worldwide as a standardized way to rate the wind force and associated wave heights. Figure 2.8 is such an example of a Beaufort Table.

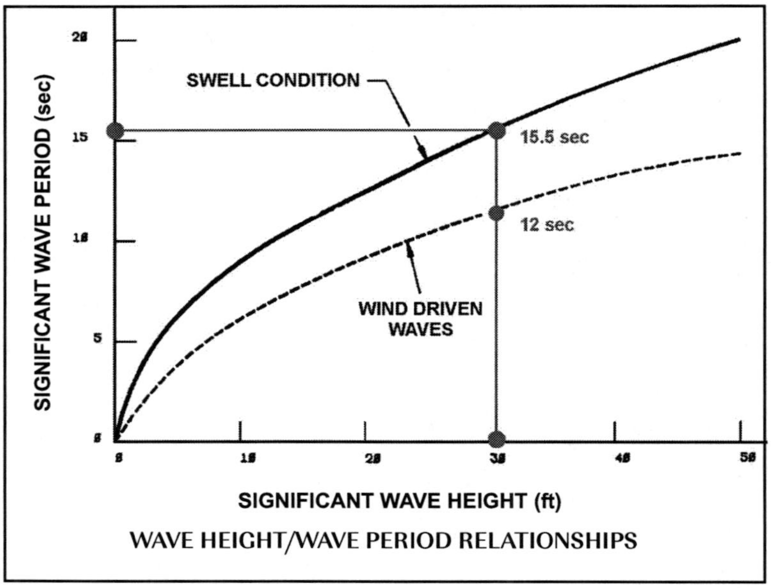

Figure 2.7 Swell and wind driven wave height versus wave period

BEAUFORT	WIND	SEA STATE AND WAVE HEIGHT LEVELS
Force 0	Less than 1 knot	Sea like a mirror
Force 1	1–3 knots	Wave height .1 m (.25 ft); ripples with appearance of scales, no foam crests.
Force 2	4–6 knots	Wave height .2–.3 m (.5–1 ft); small wavelets, crests of glassy appearance, not breaking.
Force 3	7–10 knots	Wave height .6–1 m (2–3 ft); large wavelets, crests begin to break, scattered whitecaps.
Force 4	11–16 knots	Wave height 1–1.5 m (3.5–5 ft); small waves, becoming longer, numerous whitecaps.
Force 5	17–21 knots	Wave height 2–2.5 m (6–8 ft); moderate waves, taking longer to form, many whitecaps, some spray.
Force 6	22–27 knots	Wave height 3–4 m (9.5–13 ft); larger waves, foaming, whitecaps everywhere, more spray.
Force 7	28–33 knots	Wave height 4–5.5 m (13.5–19 ft); sea heaps up, white foam from breaking waves begins to be blown in streaks along direction of wind.
Force 8	34–40 knots	Wave height 5.5–7.5 m (18–25 ft); moderately high waves of greater length, edges of crests begin to break into spindrift, foam is blown in well marked streaks.
Force 9	41–47 knots	Wave height 7–10 m (23–32 ft); high waves, sea begins to roll dense streaks of foam along wind direction, spray may reduce visibility.
Force 10	48–55 knots	Wave height 9–12.5 m (29–41 ft); very high waves with overhanging crests, sea takes white appearance as foam is blown in very dense streaks, rolling is heavy and shocklike, visibility is reduced.
Force 11	56–63 knots	Wave height 11.5–16 m (37–52 ft); exceptionally high waves, sea covered with white foam patches, visiblitiy is still more reduced.

Figure 2.8 Example of the Beaufort scale

2.12 NATURAL MOTION PERIODS

The motions of a rig depend on many factors such as:

1. Hull shape.
2. Hull dimensions.
3. Displacement.
4. Position of the center of gravity.

The natural motion period of a rig is important. As a result of the wave action, a rig oscillates around the three rotation axes with its own natural roll, pitch, and heave period. If the wave action stops, the rig will continue to oscillate with the same period but the amplitude will decrease. The friction between the rig and seawater causes a damping effect. If the wave period and the natural period of the rig are too close, the wave action will cause excessive synchronous motions. The rig designer estimates the natural period of the roll and pitch and heave with the formulas:

The formula for the natural pitch and roll period is:

$$T = 2\pi \sqrt{\frac{I}{W \cdot GM}}$$

I = the total inertia of the rig including the added mass.

The added mass is the amount of water that moves with the rig.

This amount can be considerable and depends on the shape and direction.

W = the displacement of the rig

GM = the metacentric height

The formula for the natural heave period is:

$$T = 2\pi \sqrt{\frac{W + W_a}{g \cdot \delta \cdot A}}$$

$W + W_a$ = displacement and added mass

g = acceleration of gravity

δ = density of seawater

A = waterplane area

The formulas for roll, pitch, and heave confirm that:

1. To prevent synchronous rolling and pitching, the designer will look at a natural period less than 6 seconds or more than 20 seconds.
2. Because the rig designer is restricted by minimum *GM* requirements, he looks at a natural motion period of more than 20 seconds. To accomplish this, the rig should have a combination of a large displacement, a large added mass, and a small waterplane area.
3. The heave period will be large if the rig is designed to have a small waterplane area and a large added mass. Items (2) and (3) are typical for a semisubmersible at operations draft.
4. Floating on the hull, a semisubmersible will behave like a conventional vessel. Under tow, it may therefore be necessary to ballast down to reduce excessive motions during a storm. Drilling vessels behave like conventional vessels.
5. A tender rig with a small *GM* will have a long rolling period compared to a stiff rig with a large *GM* and small rolling period.
6. Lowering the legs of a floating jackup (JU) will change the *GM* and the roll and pitch.

Figure 2.9 displays the difference of the motion characteristics between a semisubmersible and a ship-shaped drilling vessel.

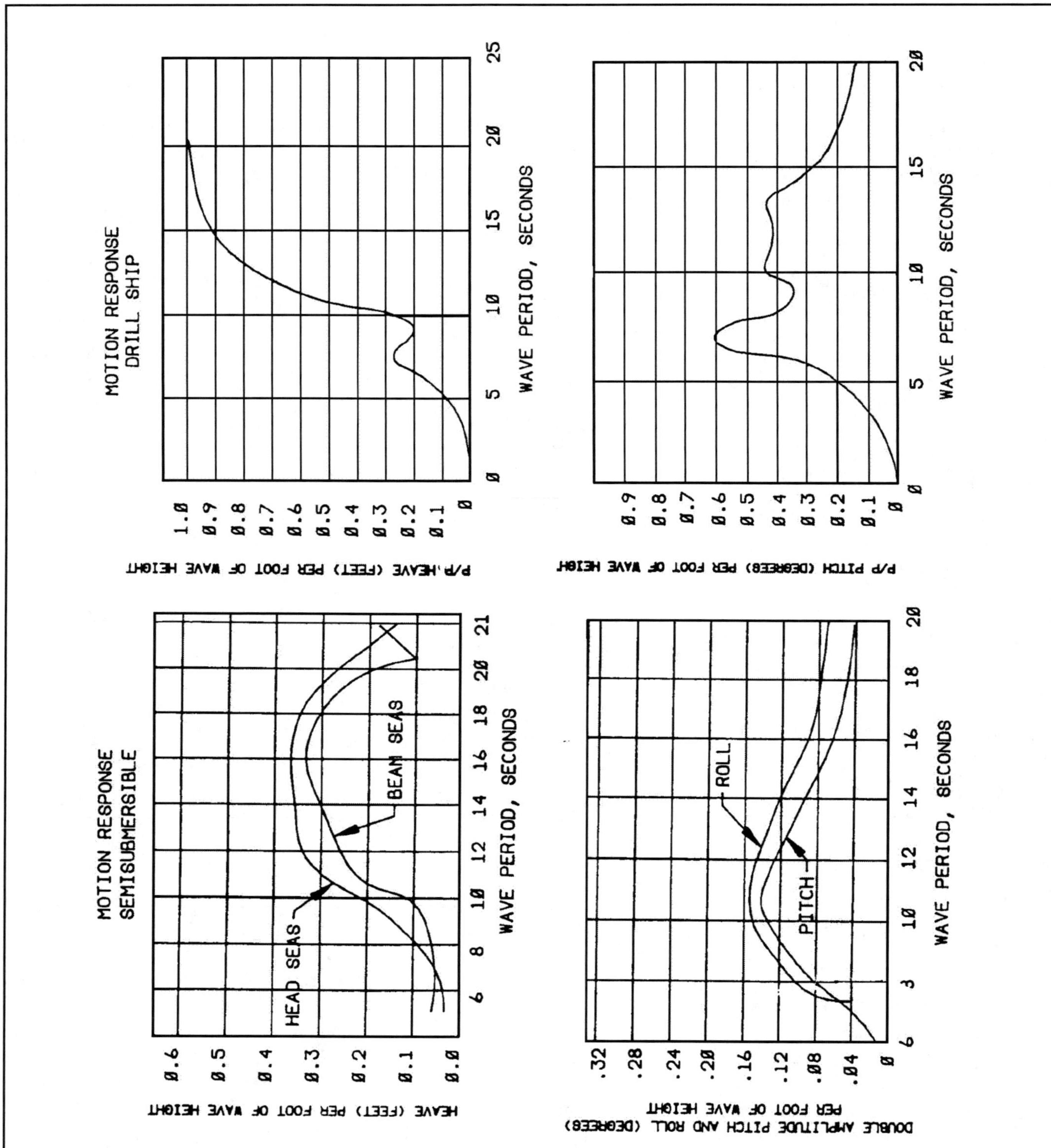

Figure 2.9 Motion characteristics—semisubmersible versus drilling vessel

Example calculation of natural heave period:

Calculate the natural heave period of a semisubmersible with four 40 ft columns and a displacement of 26,000 st. The added mass is 40% of displacement.

$g = 32.2$ ft/sec². Density seawater = 64.4 lb/ft³

Total weight = 26,000 · 2,000 = 52,000,000 lb

20,800,000 lb added mass

72,800,000 lb $W + W_a$

$$T = 2\pi\sqrt{\frac{72{,}800{,}000}{32.2 \cdot 64.4 \cdot 4\left(\frac{\pi \cdot 40^2}{4}\right)}}$$

$T = 16.6$ sec

2.13 WIND

The difference in barometric pressures between two locations generates wind. The horizontal air movement between two locations with different barometric pressures is called wind. The higher the difference in barometric pressure between the same two locations, the higher the wind speed. The friction between the water surface of the ocean and the wind causes the water surface to form ripples, which increase in height with the wind force to form waves.

As explained in section 2.10, the final result will be a maximum wave height and period corresponding with the wind speed. Areas with a constant wind, such as monsoons, are quite predictable to indicate the wind speed. In areas with very turbulent weather, such as the North Sea in the winter, the wind speed and direction vary from one moment to another. It is therefore important to define how maximum and average wind speed and wind forces will be applied to calculate the wind forces on the offshore structures.

2.14 WIND SPEED

The wind speed is computed from available records over many years which have been collected by weather stations and vessels fitted with weather recording equipment. These records are normally measured into time intervals of three seconds, one minute, and one hour. Within each interval the recorded wind speeds are averaged (fig. 2.10).

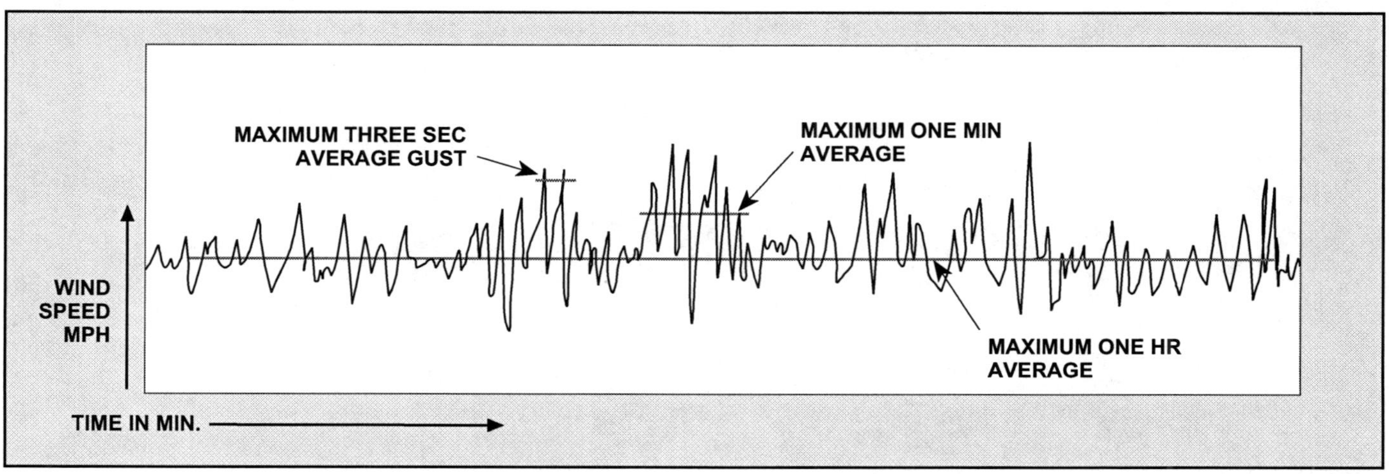

Figure 2.10 Maximum average wind speed values

1. The *three-second gust* is defined as the maximum value of the average wind speed over a three-second interval. The measurement is done at a very short interval. This represents the instantaneous wind speed. Generally, the three-second gust wind speed is used to design individual members on the open deck and not for the entire rig.
2. The *one-minute wind* is the maximum value of the average wind speed over a one-minute interval. This is the wind speed used with the maximum storm conditions for operations criteria.
3. The *one-hour wind speed* is the maximum value of the average wind speed over a one-hour interval.

It is obvious that the longer the average period, the lower the average value. The one-minute wind is less than the three-second gust and the one-hour wind is less than the one-minute wind. In ratios, the one-minute wind is 0.85 times the three-second gust and the one-hour wind is 0.86 times the one-minute wind.

2.15 WIND FORCE CALCULATION

The wind force, *F*, is obtained by wind tunnel tests or by calculation with formulas given by the classification societies. For example, ABS uses the formula:

$$F = P \cdot A$$

F = force in kg or lb

P = wind pressure in kg/m² or lb/ft²

A = area in m² or ft²

In metric → $F = 0.0623C_h \cdot C_s \cdot V^2$. For Imperial → $F = 0.00338C_h \cdot C_s \cdot V^2$

V = one-minute wind velocity in m/sec or in knots

C_h = height coefficient from Table 1

C_s = shape coefficient from Table 2

Table 2.1 Height Coefficient and Table 2.2 Shape Coefficient are examples from ABS to calculate wind forces. Other classification societies may use different tables, but the overall result will be about the same and always within the IMO criteria.

The design criteria depends on the type of rig; i.e., jackup, semisubmersible, or drilling vessel. But, in general, the calculations for environmental design limits are based on the following wind speeds:

For unrestricted offshore services

Condition	Wind speed in knots
1. Operational—Transit	
Intact	70 kt
Damaged	50 kt
2. Survival draft	
Intact	100 kt
Damaged	50 kt

For restricted services, a lower limit with a minimum wind velocity of 50 kt is used.

In some parts of the world with seasonal hurricane force winds, such as India and the Gulf of Mexico, the one-minute speed range can range from 110–130 kt.

TABLE 2.1

ABS Height Coefficient

Height Metres Over	Height Metres Not Exceeding	Height Feet Over	Height Feet Not Exceeding	(C_h) Height Coefficient
0	15.3	0	50	1.00
15.3	30.5	50	100	1.10
30.5	46.0	100	150	1.20
46.0	61.0	150	200	1.30
61.0	76.0	200	250	1.37
76.0	91.5	250	300	1.43
91.5	106.5	300	350	1.48
106.5	122.0	350	400	1.52
122.0	137.0	400	450	1.56
137.0	152.5	450	500	1.60
152.5	167.5	500	550	1.63
167.5	183.0	550	600	1.67
183.0	198.0	600	650	1.70
198.0	213.5	650	700	1.72
213.5	228.5	700	750	1.75
228.5	244.0	750	800	1.77
244.0	256.0	800	850	1.79
256		850		1.80

TABLE 2.2

ABS Shape Coefficient

Shape	(C_s) Shape Coefficient
Cylindrical shapes (all sizes)	0.5
Hull (surface type)	1.0
Deck house	1.0
Isolated structural shapes (cranes, angles, channels, beams, etc.)	1.5
Under deck areas (smooth surfaces)	1.0
Under deck areas (exposed beams and girders)	1.3
Rig derrick (each face)	1.25

In an upright condition, the wind, wave, and current forces are reasonably simple to calculate. The calculations are more complicated and require adjustments for various angles because the shape changes in respect of horizontal wind force. For a semisubmersible, the large effect of the underside of the main deck increases the wind load as the angle increases. For a jackup on location, the situation does not change but, under tow, the wind force calculation needs adjustment for the angle. For a ship-shaped unit, the wind force reduces with the cosine of the angle because the projected height of the area decreases.

The wind force is proportional to the square of the velocity; i.e., a 10% increase in wind will increase the wind force by 21%. For the same reason, the overturning moment for the 100 kt wind velocity criteria is close to double the amount of the overturning moment of the 70 kt wind criteria.

In general, the wind force is more critical than the wave force. This is particularly true for the jackup overturning moment.

To calculate the wind force, ABS applies the following standards:

1. For column-stabilized units, all columns are included. No shielding is allowed.
2. With an inclination, the exposure of the underdeck area is added.
3. The block projected area of a cluster of deck houses may be calculated as one piece with a coefficient of 1.1.
4. Isolated deck houses are to be calculated individually.
5. Open truss work, such as boom and derricks, may be approximated by taking 30% of each front and back area or 60% with the shape coefficient of block area on one side.

In some marine operations manuals, a graph allows easy calculation of the wind force with a known wind velocity in knots at a specific draft (fig. 2.11).

Example of a wind force calculation for a drilling vessel.

The beam wind force calculation with the ABS rules for a drilling vessel for a 70 kt storm condition (fig. 2.12).

2.16 CURRENT

The horizontal movement of the ocean water is called current. Tidal currents are caused by gravitational pull of the sun and the moon. The circulation of the oceans causes nontidal currents. At sea, we generally experience a combination of tidal and nontidal currents. Currents can change speed and direction at various water depths. The surface current can be opposite the subsurface current.

2.17 CURRENT LOADS

The equation to calculate the current force on the subsurface structure of a drilling rig is:

$$F_d = 0.5 \cdot C_d \cdot \rho \cdot A \cdot V^2$$

C_d = drag coefficient

ρ = density of seawater

A = projected area

V = velocity of the current

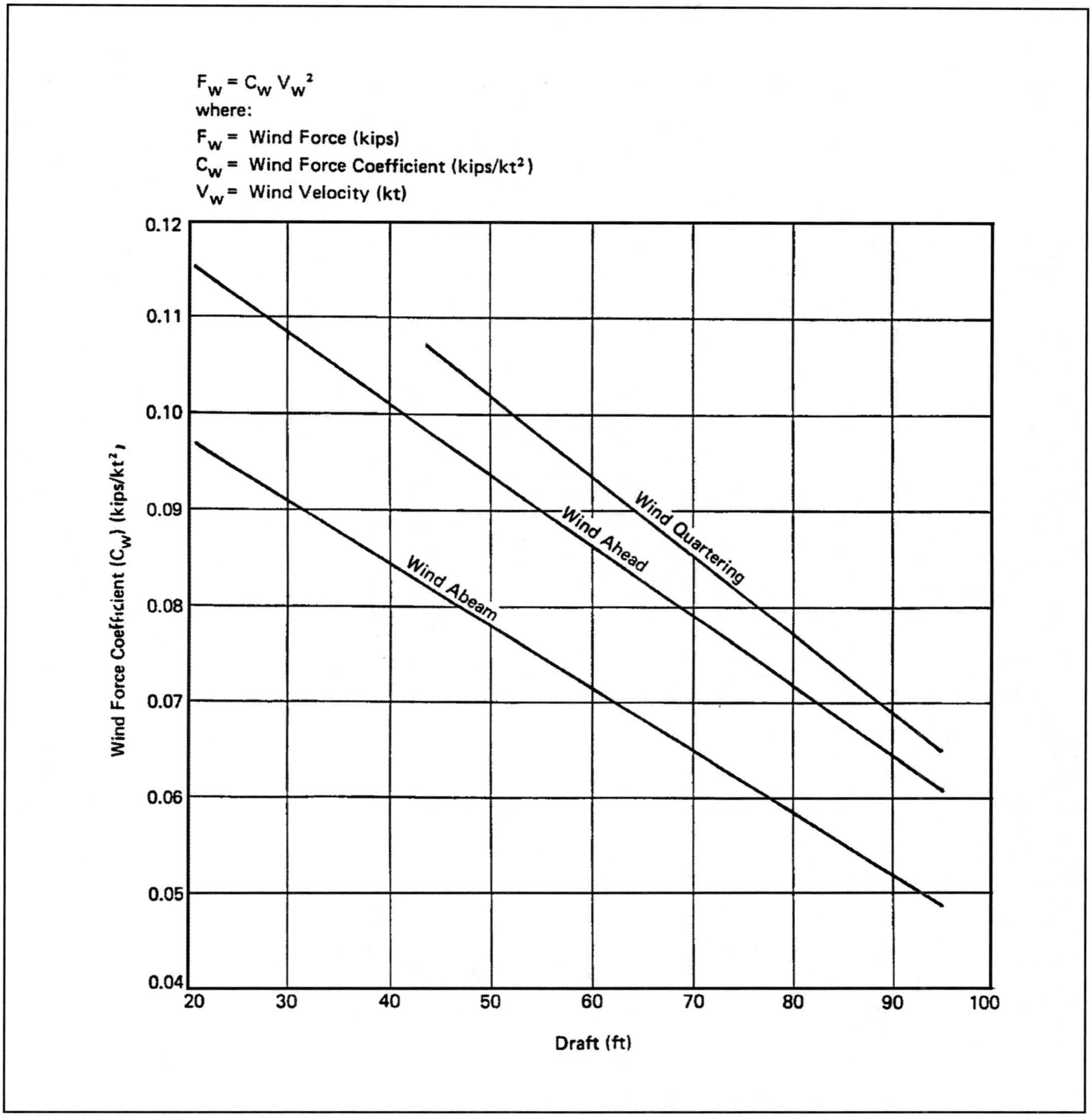

Figure 2.11 Wind force coefficient versus draft for Rig Type 2A

Hull and deck houses	Area A1	9200 ft^2
Derrick	Area A2	1800 ft^2
Derrick	Area A3	1200 ft^2
Derrick	Area A4	1020 ft^2
Derrick	Area A5	580 ft^2
2 Crane Houses	Area A6	490 ft^2
2 Crane Booms	Area A7	610 ft^2
Helideck and construction	Area A8	5900 ft^2

<u>Answers:</u>

	Table 1	Table 2		
Area	C_s	C_h	A	$C_s\ C_h\ A$
A1	1.0	1.0	9200	9200
A2	1.25	1.1	(0.6)(1800)	1485
A3	1.25	1.2	(0.6)(1200)	1080
A4	1.25	1.3	(0.6)(1020)	995
A5	1.25	1.37	(0.6)(580)	496
A6	1.0	1.0	490	490
A7	1.25	1.0	(0.6) 610	458
A8	1.3	1.0	(0.6) 5900	4602
			Total Area	18906

Total wind force F $= 0.00338 \times 18906 \times V^2$

$= 0.00338 \times 18906 \times 70^2$

$=$ 313,121 lbs

$=$ 156.6 ST

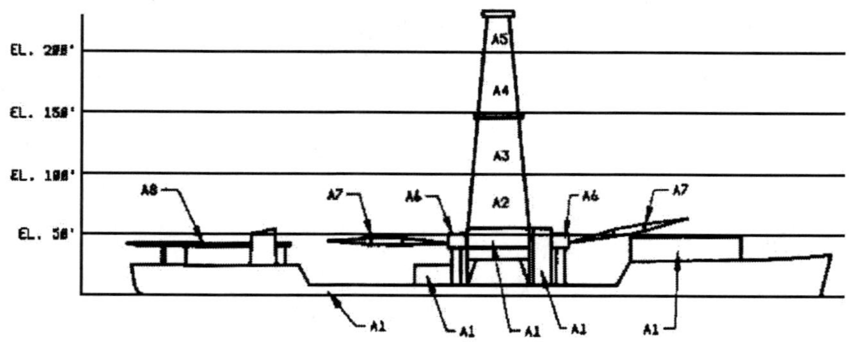

Figure 2.12 Example wind force calculation for a drilling vessel

This formula is similar to the formula discussed for the wave drag force. The current will only generate drag forces. The combined formula for wave and current is therefore:

$$F_d = 0.5 \cdot C_d \cdot \rho \cdot A \cdot (V_w + V_c)^2$$

V_w = the horizontal velocity of the water particles due to wave action

V_c = the current velocity

The above formula shows that wave force is the function of the square of the sum of the velocities of wave and current $(V_w + V_c)^2$. This is larger than $(V^2_w + V^2_c)$. The effect of the current is much more than one would expect.

Like for wave and wind on some rigs, the rig's marine operations manual (MOM) shows a graph with the current force coefficient which enables a quick calculation of the current force at a certain draft if the current velocity is known (figs. 2.5, 2.11, and 2.13).

The summation of the forces from the graphs wave, wind, and current (figs. 2.5, 2.11, and 2.12) provides for a quick calculation of the total environmental forces acting on the rig at a specific draft.

2.18 EXAMPLES FROM MOM OF APPROVED DESIGN CONDITIONS

Based on the explanation in this chapter, examples from the MOM with the maximum design criteria are shown in Figure 2.14 for Rig Type 1 and Figure 2.15 for Rig Type 4.

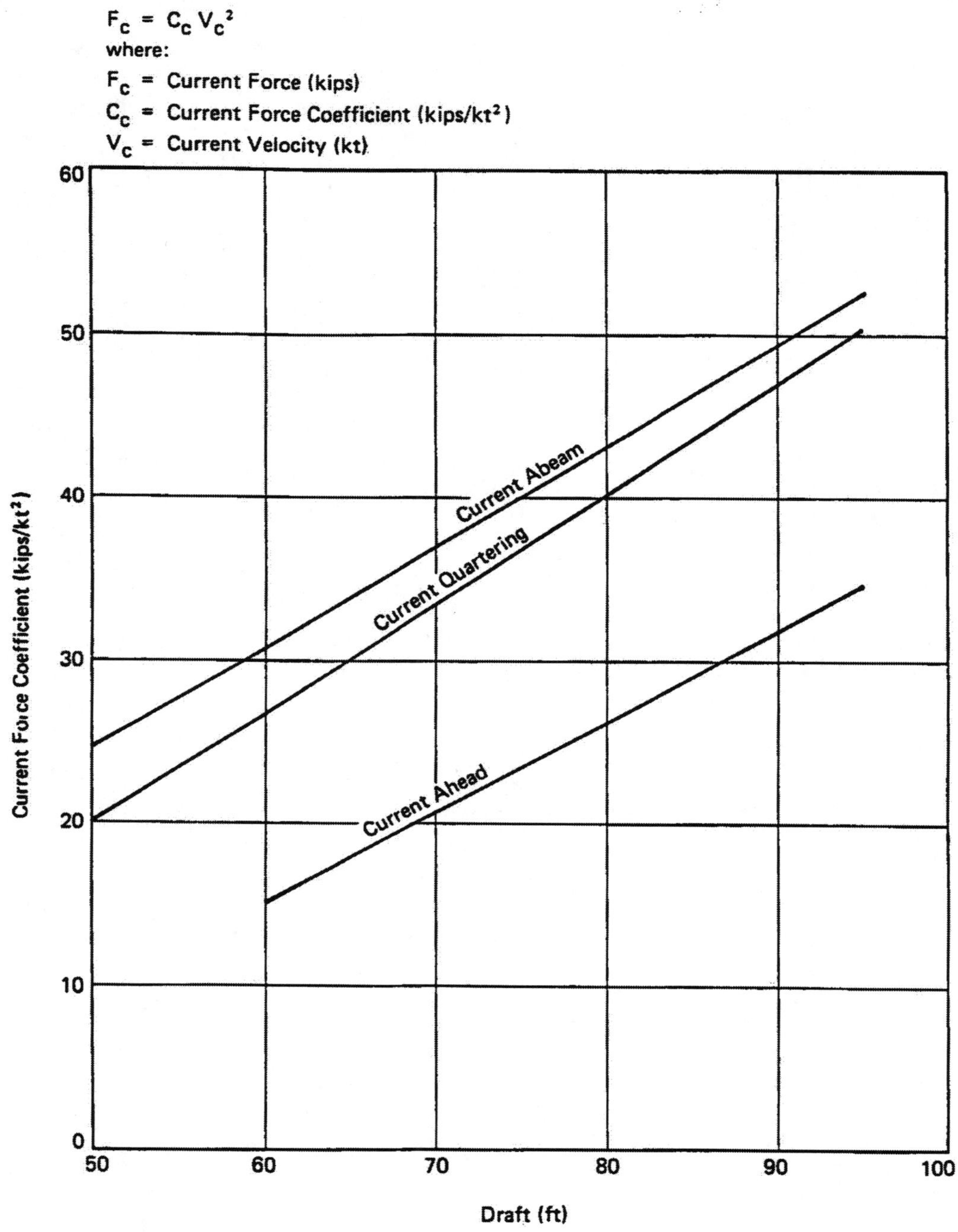

Figure 2.13 Current force coefficient versus draft Rig Type 2A

The rig is approved for the following maximum environmental (design) conditions:

	Operational	Survival	Transit
Wave Height m (ft)	6.00 (19.69)	12.00 (39.37)	3.00 (9.84)
Wave Period, s	10.00	12.00	8.00
Wind Speed, kn (m/s)	70.00 (36.00)	100.00 (51.40)	40.0 (20.60)
Current Speed, kn (m/s)	1.80 (3.50)	3.00 (5.82)	N/A

The ABS approved design steel temperatures is 0°C for water and air, respectively.

The rig when operated in accordance with guidance given here in complies with the stability criteria contained in:

- American Bureau of Shipping "Rules and Regulations for the Building and Classing of Mobile Offshore Drilling Units" (1998)
- The 1989 IMO Code, IMO Resolution A649 (XVI) as amended 1991.

Figure 2.14 Design criteria Rig Type 1

RIG TYPE 4
SECTION 2 – LIMITS OF SERVICE

Table 2-3 – Environmental and Motion Limits

	Loading Condition*				
	1	2	3	4	5
Environmental Description					
Wind, knots (1 min sustained)	40	60	60	100	50
Sigificant Wave Height, m	5.5	9.1	7.6	15.2	0
Heading Off Bow, degrees	±30	±30	±30	±30	±90
Motions Description					
Static List or Trim, degrees	0	3	3	5	17
Pitch, degrees (double amplitude)	2.96	8.09	5.54	15.46	0.0
Pitch Period, seconds (double amplitude)	8.9	12.1	12.1	12.9	0.0
Roll, degrees (double amplitude)	2.05	1.69	2.01	11.75	0.0
Roll Period, seconds (double amplitude)	17.0	17.4	17.4	17.5	0.0
Heave, m (double amplitude)	1.94	4.71	3.86	15.56	0.0
Heave Period, seconds (double amplitude)	9.7	12.8	12.8	14.6	0.0
Drilling Loads Description					
Hook/Rotary Load – Primary, kips	2,000	1,400	1,800	0	0
Hook/Rotary Load – Secondary, kips	2,000	1,200	1,400	0	0
Overhead & Travelling Load – Primary, kips	100	200	100	200	200
Overhead & Travelling Load – Second., kips	100	200	100	200	200
Crown Load – Primary, kips	2,100	1,600	1,900	200	200
Crown Load – Secondary, kips	2,100	1,400	1,500	200	200
Drill Pipe Setback, kips (x 2)	1,250	1,250	1,250	1,250	1,250
Casing Setback, kips	1,000	1,000	1,000	0	1,000
Tensioner Load, kips (x 2)	3,200	2,500	2,500	4,120	2,500
Travelling Equip. Elev., ft Above Drill Floor	110	110	110	30	30

*Description of Loading Conditions:

1. Maximum hook load condition – special operating condition.
2. Maximum operating condition – drilling.
3. Maximum operating condition – tripping or running casing/riser.
4. Maximum storm condition.
5. Damaged condition.

Figure 2.15 Design criteria Rig Type 4

EXERCISE 1
Chapter 2

1. Name the rig motions about:
   a. The longitudinal axis.
   b. The vertical axis.
   c. The transverse axis.

2. Of the above,
   a. Which motions are rotational?
   b. Which motions are translational?

3. Determine the wave crest velocity in ft/sec of an 11 sec regular wave with a wave length of 400 ft.

4. What is the wave velocity in kt? (1 kt = 6,076 ft/hour)

5. What is the frequency of the wave from question 3?

6. Calculate the wave velocity in ft/sec for a 42 ft high (double amplitude) regular wave with an angular velocity of .495 rad/sec.

7. What is the answer from question 6 in kt?

8. What is the period in seconds of the wave from question 6?

9. Determine from Figure 2.7 the maximum wave height for a wind driven wave with a 10 sec significant period.

EXERCISE 2
Chapter 2

1. From Figure 2.9.

   a. What is the heave for a semisubmersible in 25 ft seas from the bow with a period of 11 sec?
   b. What is the pitch in degrees for a drillship in 8 ft seas with a period of 5 sec?

2. If you were designing a semisubmersible and want to increase its natural heave period, what would you do?

   a. Increase the diameter of the columns.
   b. Increase the displacement.
   c. Decrease the width of the pontoons to reduce the added mass.

   Explain your answer.

3. Calculate the total wind force on a horizontal flare boom 125 ft long consisting of open truss work. The boom is rectangular in cross section 8 ft each side. Wind speed is 75 kt blowing perpendicular to the boom (broadside). Distance from the waterline to center area of the boom is 55 ft.

4. What will happen if a drilling vessel experiences a sea state where the wave period is close to the vessel's natural period in roll? What action can you take?

CHAPTER 3
Hydrostatic Properties and the Daily Stability Calculations

3.1 INTRODUCTION

During the design and building stage, the naval architect calculates, with special computer generated programs, the rig's hydrostatic data.

The details of the hydrostatic data are tabulated in the hydrostatic properties. In addition, curves of stability are analyzed to confirm that the stability criteria is in accordance with the standards of the IMO, classification society, flag regulations, and local regulations. The information from the hydrostatic properties is used to perform the daily stability calculations.

Separate information not part of the hydrostatic properties may be:

1. tank capacity curves or tables.
2. capacity, weight, and moment of flooded compartments.
3. free surface effect of hull and ballast tanks.
4. the lightship information and history.
5. the allowable VCG values.

3.2 HYDROSTATIC PROPERTY INFORMATION

Tables 3.1, 3.2, and 3.3 are typical examples of hydrostatic tables of two semisubmersibles and one drilling vessel.

The information in Imperial or metric system includes:

1. Draft in ft or m.
2. Displacement in st or mt (Displ).
3. Vertical center of buoyancy in ft or m (VCB or KB).
4. Longitudinal center of buoyancy in ft or m (LCB).
5. Transverse center of buoyancy in ft or m (TCB).
6. Waterplane area in square ft or square m (WPA).
7. Short tons per inch (TPI) or tonnes per cm (TPC).
8. Longitudinal center of flotation in ft or m (LCF).
9. Transverse center of flotation in ft or m (TCF).
10. The height of the longitudinal metacenter in ft or m (KM_L).
11. The height of the transverse metacenter in ft or m (KM_T).
12. The moment to change the trim one degree or one cm (MT1° or MCT).
13. The moment to change the heel one degree or one cm (MH1° or MCH).

3.3 DRAFT

The draft is calculated at intervals of feet and metres or parts thereof.

TABLE 3.1
Example Rig Type 2A Hydrostatic Properties
Imperial system

RIG TYPE 2A
HYDROSTATIC PROPERTIES
(Page 3 of 4)

DRAFT ft	DISPL ST	KB ft	TP1" ST	KML ft	KMT ft	MH1deg (ft-ST) VCG=50	MH1deg (ft-ST) VCG=60	MH1deg (ft-ST) VCG=65	MT1deg (ft-ST) VCG=50	MT1deg (ft-ST) VCG=60	MT1deg (ft-ST) VCG=65
61.0	24765	18.00	11.14	69.42	67.43	7533	3211	1050	8393	4071	1910
62.0	24900	18.23	11.14	69.38	67.37	7548	3203	1030	8422	4076	1903
63.0	25030	18.47	11.14	69.34	67.31	7562	3193	1009	8448	4080	1896
64.0	25165	18.71	11.14	69.31	67.27	7585	3193	997	8481	4089	1893
65.0	25300	18.95	11.14	69.29	67.24	7612	3197	989	8517	4102	1894
66.0	25430	19.19	11.14	69.27	67.20	7634	3195	976	8552	4114	1895
67.0	25565	19.44	11.14	69.25	67.17	7661	3199	968	8589	4127	1896
68.0	25700	19.69	11.14	69.24	67.15	7692	3207	964	8630	4144	1902
69.0	25835	19.94	11.14	69.24	67.13	7724	3215	960	8675	4166	1912
70.0	25965	20.20	11.14	69.24	67.12	7758	3226	961	8719	4187	1921
70.5	26035	20.33	11.14	69.24	67.11	7774	3231	959	8742	4198	1927
71.0	26100	20.46	11.14	69.25	67.11	7794	3239	961	8769	4213	1936
71.5	26170	20.58	11.14	69.25	67.11	7815	3247	964	8792	4225	1941
72.0	26235	20.72	11.14	69.26	67.11	7834	3255	966	8818	4240	1950
72.5	26300	20.85	11.14	69.27	67.11	7853	3263	968	8845	4255	1960
73.0	26370	20.98	11.14	69.28	67.11	7874	3272	971	8873	4271	1970
73.5	26435	21.11	11.14	69.29	67.12	7898	3285	978	8900	4286	1979
74.0	26500	21.24	11.14	69.30	67.12	7918	3293	980	8926	4301	1989
74.5	26570	21.38	11.14	69.31	67.13	7943	3306	988	8954	4317	1999
75.0	26635	21.51	11.14	69.32	67.14	7967	3319	995	8981	4332	2008
75.5	26705	21.65	11.14	69.34	67.15	7993	3332	1002	9014	4353	2023
76.0	26770	21.78	11.14	69.35	67.16	8017	3345	1009	9040	4368	2032
76.5	26835	21.91	11.14	69.37	67.17	8041	3358	1016	9072	4388	2047
77.0	26905	22.05	11.14	69.39	67.18	8067	3371	1024	9105	4409	2061
77.5	26970	22.19	11.14	69.41	67.20	8096	3389	1036	9136	4429	2076
78.0	27035	22.33	11.14	69.43	67.21	8120	3402	1043	9168	4449	2090
78.5	27105	22.46	11.14	69.45	67.23	8151	3420	1055	9201	4470	2105
79.0	27170	22.60	11.14	69.47	67.25	8180	3438	1067	9232	4491	2120
79.5	27235	22.74	11.14	69.50	67.27	8209	3456	1079	9269	4516	2139
80.0	27305	22.88	11.14	69.52	67.28	8235	3469	1087	9302	4537	2154
80.5	27370	23.02	11.14	69.55	67.31	8269	3492	1103	9338	4562	2173
81.0	27440	23.16	11.14	69.58	67.33	8299	3510	1116	9377	4588	2193
81.5	27505	23.30	11.14	69.60	67.35	8328	3528	1128	9409	4608	2208
82.0	27570	23.44	11.14	69.63	67.38	8363	3551	1145	9445	4634	2228
82.5	27640	23.59	11.14	69.66	67.40	8393	3570	1158	9484	4660	2248
83.0	27705	23.73	11.14	69.70	67.43	8428	3593	1175	9525	4690	2273
83.5	27770	23.87	11.14	69.73	67.46	8462	3616	1192	9562	4716	2292
84.0	27840	24.02	11.14	69.76	67.49	8498	3639	1210	9601	4742	2313
84.5	27905	24.16	11.14	69.80	67.52	8532	3662	1227	9643	4773	2338
85.0	27975	24.31	11.14	69.83	67.55	8568	3686	1245	9682	4799	2358

TABLE 3.2
Example Hydrostatic Properties Rig Type 1
Metric system

HYDROSTATIC PROPERTIES
Metric (Sheet 6 of 8)

Draft (m)	Displt. (tonnes)	VCB (m)	LCB (m)	TCB (m)	WPA (m^2)	TPC (t/cm)	LCF (m)	TCF (m)	KM_L (m)	KM_T (m)	MCT (6) (t-m/cm)	MCH (6) (t-m/cm)
19.00	33,766	6.77	52.97	0.00	684	7.04	52.00	0.00	23	23	74	112
19.10	33,836	6.79	52.97	0.00	684	7.04	52.00	0.00	23	23	75	112
19.20	33,907	6.82	52.96	0.00	684	7.04	52.00	0.00	23	23	75	112
19.30	33,977	6.85	52.96	0.00	684	7.04	52.00	0.00	23	23	75	113
19.40	34,048	6.87	52.96	0.00	684	7.04	52.00	0.00	23	23	75	113
19.50	34,118	6.90	52.96	0.00	684	7.04	52.00	0.00	23	23	75	113
19.60	34,189	6.92	52.96	0.00	684	7.04	52.00	0.00	23	23	75	113
19.70	34,259	6.95	52.95	0.00	684	7.04	52.00	0.00	23	23	75	113
19.80	34,330	6.98	52.95	0.00	684	7.04	52.00	0.00	23	23	75	114
19.90	34,400	7.00	52.95	0.00	684	7.04	52.00	0.00	23	23	76	114
20.00	34,470	7.03	52.95	0.00	684	7.04	52.00	0.00	23	23	76	114
20.10	34,541	7.06	52.95	0.00	684	7.04	52.00	0.00	23	23	76	114
20.20	34,611	7.08	52.94	0.00	684	7.04	52.00	0.00	23	23	76	114
20.30	34,682	7.11	52.94	0.00	684	7.04	52.00	0.00	23	23	76	115
20.40	34,752	7.14	52.94	0.00	684	7.04	52.00	0.00	23	23	76	115
20.50	34,823	7.16	52.94	0.00	684	7.04	52.00	0.00	23	23	76	115
20.60	34,893	7.19	52.94	0.00	683	7.04	52.03	0.00	23	23	77	115
20.70	34,963	7.22	52.94	0.00	682	7.02	52.08	0.00	23	23	77	115
20.80	35,033	7.24	52.93	0.00	681	7.01	52.13	0.00	23	23	77	115
20.90	35,103	7.27	52.93	0.00	679	6.99	52.18	0.00	23	23	77	116
21.00	35,173	7.30	52.93	0.00	678	6.98	52.23	0.00	23	23	77	116
21.10	35,243	7.33	52.93	0.00	676	6.97	52.28	0.00	23	23	77	116
21.20	35,313	7.35	52.93	0.00	675	6.95	52.33	0.00	23	23	77	116
21.30	35,382	7.38	52.93	0.00	674	6.94	52.38	0.00	23	23	77	116
21.40	35,451	7.41	52.93	0.00	672	6.92	52.44	0.00	23	23	77	116
21.50	35,521	7.43	52.93	0.00	671	6.91	52.49	0.00	23	23	77	116
21.60	35,590	7.46	52.92	0.00	669	6.89	52.55	0.00	23	23	77	116
21.70	35,659	7.49	52.92	0.00	668	6.88	52.60	0.00	23	23	77	117
21.80	35,727	7.52	52.92	0.00	666	6.86	52.66	0.00	23	23	77	117
21.90	35,796	7.54	52.92	0.00	665	6.85	52.72	0.00	23	23	77	117
22.00	35,864	7.57	52.92	0.00	663	6.83	52.78	0.00	23	22	77	117
22.10	35,932	7.60	52.92	0.00	662	6.82	52.84	0.00	23	22	77	117
22.20	36,001	7.63	52.92	0.00	660	6.80	52.90	0.00	23	22	77	117
22.30	36,068	7.65	52.92	0.00	659	6.78	52.96	0.00	23	22	77	117
22.40	36,136	7.68	52.92	0.00	657	6.77	53.02	0.00	23	22	77	117
22.50	36,204	7.71	52.92	0.00	656	6.75	53.08	0.00	23	22	77	117
22.60	36,271	7.74	52.92	0.00	654	6.74	53.15	0.00	23	22	77	118
22.70	36,339	7.76	52.92	0.00	652	6.72	53.21	0.00	23	22	77	118
22.80	36,406	7.79	52.92	0.00	651	6.70	53.28	0.00	23	22	77	118
22.90	36,473	7.82	52.92	0.00	649	6.68	53.35	0.00	23	22	78	118

TABLE 3.3
Example Hydrostatic Properties Rig Type 4
Metric system

FORAN SYSTEM - MODULE HYDROS VERSION 30 RESULTS PAGE (
DATE - 98.05.
CUSTOMER - SONAT O. DR. INC FN
DESCRIPTION OF SHIP - D. P. DRILLING UNIT a2

HYDROSTATIC CHARACTERISTICS

AFT PP DRAFT (TA) M	FORE PP DRAFT (TF) M	DISPLA. WITH APPEND. (DISFA) TM	SHIP MOULDED VOLUME (DISV) M3	L.C.B WITH APPEND. (XCBA)* M	C.OF B. ABOVE B/LINE (KB) M	TOTAL BLOCK COEFF. (CB) -	MOMENT FOR TRI 1 CM (MTC) MTM
12.000	12.000	93022.	90301.	114.693	6.177	0.8270	1631.
12.200	12.200	94728.	91964.	114.712	6.283	0.8284	1645.
12.400	12.400	96441.	93632.	114.733	6.390	0.8298	1660.
12.600	12.600	98159.	95307.	114.756	6.498	0.8312	1674.
12.800	12.800	99884.	96988.	114.781	6.605	0.8326	1688.
13.000	13.000	101615.	98675.	114.807	6.712	0.8340	1700.
13.200	13.200	103351.	100366.	114.833	6.820	0.8355	1713.
13.400	13.400	105092.	102063.	114.860	6.927	0.8369	1724.
13.600	13.600	106838.	103764.	114.887	7.035	0.8383	1735.
13.800	13.800	108589.	105470.	114.913	7.142	0.8397	1746.
14.000	14.000	110344.	107181.	114.940	7.250	0.8411	1756.
14.200	14.200	112103.	108895.	114.965	7.358	0.8425	1765.
14.400	14.400	113864.	110612.	114.989	7.465	0.8438	1771.
14.600	14.600	115629.	112331.	115.011	7.573	0.8452	1778.
14.800	14.800	117396.	114054.	115.031	7.680	0.8465	1785.
15.000	15.000	119167.	115779.	115.050	7.787	0.8479	1791.
15.200	15.200	120940.	117507.	115.067	7.895	0.8492	1798.
15.400	15.400	122716.	119238.	115.083	8.002	0.8505	1805.
15.600	15.600	124495.	120971.	115.098	8.109	0.8518	1812.
15.800	15.800	126269.	122701.	115.118	8.216	0.8530	1807.
16.000	16.000	128061.	124447.	115.124	8.324	0.8543	1826.
16.200	16.200	129849.	126189.	115.135	8.431	0.8555	1833.
16.400	16.400	131640.	127934.	115.145	8.538	0.8568	1840.
16.600	16.600	133433.	129683.	115.154	8.645	0.8580	1848.
16.800	16.800	135230.	131433.	115.162	8.752	0.8592	1855.
17.000	17.000	137030.	133187.	115.169	8.859	0.8604	1862.
17.200	17.200	138824.	134936.	115.168	8.966	0.8615	1869.
17.400	17.400	140629.	136695.	115.174	9.073	0.8627	1877.
17.600	17.600	142437.	138457.	115.178	9.180	0.8639	1885.
17.800	17.800	144248.	140222.	115.181	9.287	0.8651	1892.
18.000	18.000	146062.	141989.	115.184	9.394	0.8662	1900.

FORAN SYSTEM - MODULE HYDROS VERSION 30 RESULTS PAGE 0
DATE - 98.05.
CUSTOMER - SONAT O. DR. INC FN.
DESCRIPTION OF SHIP - D. P. DRILLING UNIT a2

AFT PP DRAFT (TA) M	FORE PP DRAFT (TF) M	TONS PER CM (TCI) TM/CM	WATER-LINE C.O.G. (XCF)* M	METAC. OVER BASE/L (KM) M	TRANSV. METACE. RADIUS (ZBM) M	LONGIT. METACE. RADIUS (ZBML) M
12.000	12.000	85.11	115.669	16.525	10.349	420.84
12.200	12.200	85.45	115.833	16.525	10.242	417.01
12.400	12.400	85.77	115.982	16.527	10.137	413.21
12.600	12.600	86.08	116.118	16.530	10.032	409.42
12.800	12.800	86.38	116.234	16.533	9.928	405.60
13.000	13.000	86.65	116.331	16.536	9.823	401.72
13.200	13.200	86.92	116.413	16.538	9.718	397.81
13.400	13.400	87.17	116.471	16.540	9.613	393.88
13.600	13.600	87.40	116.517	16.543	9.509	389.94
13.800	13.800	87.63	116.551	16.547	9.405	386.01
14.000	14.000	87.85	116.574	16.552	9.302	382.13
14.200	14.200	88.01	116.538	16.544	9.187	377.87
14.400	14.400	88.15	116.463	16.536	9.071	373.41
14.600	14.600	88.29	116.392	16.531	8.959	369.10
14.800	14.800	88.42	116.321	16.530	8.850	364.93
15.000	15.000	88.56	116.251	16.531	8.744	360.87
15.200	15.200	88.70	116.182	16.536	8.641	356.95
15.400	15.400	88.85	116.114	16.544	8.542	353.14
15.600	15.600	88.99	116.047	16.554	8.445	349.44
15.800	15.800	88.92	116.340	16.551	8.335	343.54
16.000	16.000	89.27	115.916	16.583	8.260	342.34
16.200	16.200	89.42	115.851	16.601	8.170	338.94
16.400	16.400	89.56	115.788	16.622	8.084	335.62
16.600	16.600	89.70	115.725	16.645	8.000	332.39
16.800	16.800	89.84	115.664	16.670	7.917	329.24
17.000	17.000	89.99	115.604	16.697	7.837	326.18
17.200	17.200	90.13	115.550	16.726	7.760	323.26
17.400	17.400	90.29	115.503	16.757	7.684	320.45
17.600	17.600	90.43	115.450	16.790	7.609	317.65
17.800	17.800	90.58	115.396	16.825	7.537	314.90
18.000	18.000	90.73	115.341	16.861	7.467	312.21

- NOTE.- SYMBOL ( )* MEANS REFERRED TO LBP/2. (POSITIVE AFT)

3.4 DISPLACEMENT

As explained in Chapter 1, displacement is the weight in tonnes or short tonnes of the volume of displaced water. The draft and displacement are interrelated.

The draft and displacement figures are based on salt water with a density of 1.025 t/m³. A draft or displacement correction is required for a different density.

3.5 KB AND VCB

The values for the distance between the keel and the vertical center of buoyancy are calculated by the naval architect for each interval of the draft. The value KB depends on the underwater shape and gradually increases with the draft.

3.6 LCB AND TCB

The coordination of B with respect to the longitudinal and transverse direction is expressed in the horizontal plane as LCB and TCB. The reference lines can be the centerlines of the rig or the forward or aft perpendicular. Plus (+) and minus (-) signs are used as appropriate.

3.7 THE WATERPLANE AREA (WPA)

The waterplane area (WPA) is important for the stability behavior of the rig and the TPI and TPC calculations. Not all rigs have the WPA displayed in the hydrostatic properties.

Considerable changes in the waterplane area are noticeable when the pontoons or tubulars pass through the waterplane. As indicated in Table 3.4 in the gray marked section, the waterplane changes from pontoon to column size on Rig Type 1.

Some calculations for waterplane areas common to the drilling rigs are shown in Figure 3.1.

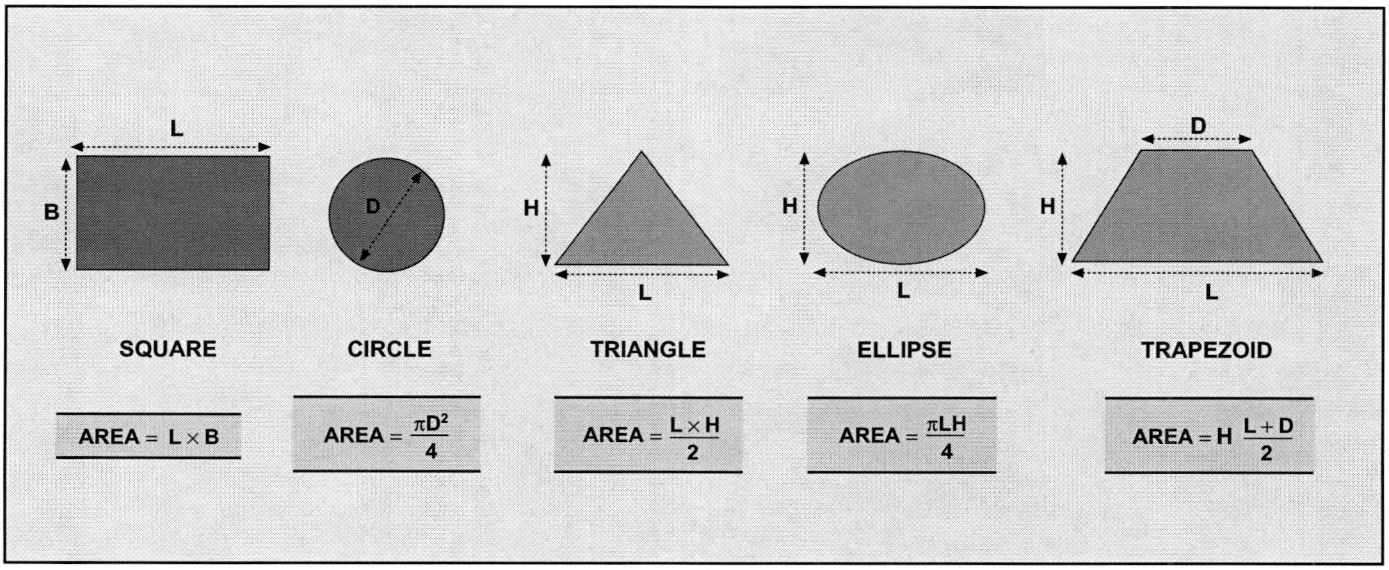

Figure 3.1 Waterplane areas

Example. Waterplane area

What is the waterplane area of a four-column semisubmersible at column draft? The columns are circles with a diameter of 40 ft.

$$\text{Four columns each area} = \frac{\pi D^2}{4} = \frac{3.14 \times 40^2}{4} \text{ ft} = 1{,}256 \text{ ft}^2$$

$$\text{Total waterplane area} = 4 \times 1{,}256 \text{ ft}^2 = 5{,}024 \text{ ft}^2$$

TABLE 3.4

Change in waterplane, TPC, KM, MCT, and MCH

Draft	Displ	VCB	LCB	TCB	WPA	TPC	LCF	TCF	KML	KMT	MCT	MCH
9.80	26,309	4.71	53.15	0.00	1,480	15.24	51.18	0.00	81	49	201	187
9.90	26,461	4.74	53.13	0.00	1,476	15.20	51.20	0.00	80	49	200	187
10.00	26,613	4.77	53.12	0.00	1,184	12.20	53.38	0.39	57	40	142	153
10.10	26,712	4.79	53.12	0.00	954	9.83	52.68	0.00	39	33	99	128
10.20	26,809	4.81	53.12	0.00	934	9.62	52.63	0.00	38	32	96	125

3.8 SHORT TONS PER IN.—TONNES PER CM (TPI–TPC)

The calculation for the short ton per in. (TPI) or tonnes per cm (TPC) is explained in Chapter 1. Table 3.4 shows the effect of the change in waterplane area going from pontoon draft to column draft.

Example. TPI

What is the TPI for the waterplane from previous *example*?

From Chapter 1 we learned:

$$\text{TPI} = \frac{\text{Waterplane area ft}^2}{375} = \frac{5{,}024}{375}$$

$$\text{TPI} = 13.40 \text{ st/in.}$$

Example. COD-TPC (See Table 3.2)

The displacement is 34,376 t. The ballast tanks are filled with 250 tonnes. What is the new draft?

First interpolate for the draft.

Draft 19.80 m → Displacement = 34,330 t Draft 19.90 m → Displacement = 34,400 t

$$\text{Difference in Displacement} = 70 \text{ t} \rightarrow \text{Proportional } \frac{46 \text{ t}}{70 \text{ t}} \times 10 \text{ cm} = 7 \text{ cm (rounded off)}$$

Draft at 34,376 t = 19.80 m + 0.07 m = 19.87 m

$$\text{New Displacement } 34{,}376 \text{ t} + 250 \text{ t} = 34{,}626 \text{ t} \rightarrow \text{TPC} = 7.04 \text{ TPC} \rightarrow \frac{250}{7.04} = 0.36 \text{ m (rounded off)}$$

New Draft is 19.87 m + 0.36 m = 20.23 m

Example. TPI

The draft of the rig in Table 3.1 is 75.3 ft. The draft needs to increase 4.7 ft. Decide how much short ton needs to be loaded.

Increase in draft is 4.7 ft

To convert to ft and in. 0.7 ft = 0.7 × 12" = 8.4"

4'8.4" = 56.4"

TPI = 11.14

Increase in draft = 11.14 TPI × 56.4" = 628.3 st

Old displacement interpolated = $26{,}635 \text{ st} + \frac{0.3}{0.5}(26{,}705 \text{ st} - 26{,}635 \text{ st}) = 26{,}677 \text{ st}$

New displacement = 26,677 st + 628.3 st = 27,305.3 st

New draft = 75.3 + 4.7 = 80 ft

3.9 LONGITUDINAL AND TRANSVERSE CENTER OF FLOTATION

On some rigs, the longitudinal center of flotation (LCF) and transverse center of flotation (TCF) are not included in the hydrostatic properties. For symmetrical waterplanes, LCF and TCF are on the centerlines of the rig.

3.10 THE HEIGHT OF THE LONGITUDINAL METACENTER AND TRANSVERSE METACENTER—KM_L AND KM_T

As explained in Chapter 1, the distance between K and M is derived from KM = KB + BM. The longitudinal and vertical values for KM are different because of the effect of the waterplane area and moment of inertia. On a semisubmersible, the difference between KM_L and KM_T is less obvious as on a vessel-shaped drilling rig. Notice in Table 3.3 the very large difference between KM_L and KM_T by looking at the difference between the transverse and longitudinal BM values.

Example.

Use Table 3.3. The draft is 12.00 m. What is the value of KM_T and KM_L?

For 12 m draft → KM = KB + BM = 6.177 m + 10.349 m = 16.525 m or if you look direct to KM

KM_T for 12 m draft = 16.525 m

For 12 m draft KM_L = $KB + BM_L$ = 6.177 m + 420.84 m = 427.017 m

3.11 MOMENT TO HEEL OR TRIM 1° (MH1°-MT1°)—MOMENT TO TRIM OR HEEL 1 CM (MCT-MCH)

During the day-to-day operations, loads are shifted, discharged, and/or loaded continuously. Each load change will not only affect the position of G but also the draft, trim, and heel. The responsibility of the barge control operator is to maintain the heel and trim at 0°. In order to maintain the rig at even keel, ballast is used with a set of forward and aft tanks (trim and heel tanks) to counteract the load changes.

The calculation for moment to trim and heel/trim 1° is explained in Chapter 1. The hydrostatic properties give the values for the moments to heel or trim the rig 1° or 1 cm or 1". This enables us to find a quick answer on the effect of load changes on the heel and trim and the required counteraction.

For asymmetric waterplanes, LCF and TCF are not positioned on the centerlines of the rig. In that case, the two parts of the trim need to be divided proportional to the offset of the center of flotation with relation to the total length. With LCF amidships, change of trim forward and aft are equal.

Notice the change in MCT and MCH of Rig Type 1 in Table 3.4 when the waterplane changes from pontoon to columns.

Example. MT1"

Like for MCT, it is possible that the hydrostatic properties are based on MT1". The calculation is the same as for MCT/ MCH.

It is easy to change the angle θ of trim or heel in degrees into trim or heel in inches or cm.

$$\tan\theta = \frac{\text{trim}}{L}$$

$$L = \text{the length or breadth of the rig}$$

$$\text{trim} = L \times \tan\theta \text{ or } 0.5 \text{ trim} - 0.5\, L \tan\theta$$

A weight of 165 tons is loaded and placed 10 ft in front of the COF. The initial draft is 69 ft forward and 71 ft aft. MT1" is 410 st•ft/in. TP1" is 15 st/in. Calculate the new draft and trim.

	Draft Fwd	Aft
Initial Draft	69'00"	71'00"
Change of mean draft $\frac{165 \text{ st}}{15 \text{ st/in.}}$ =	+ 11	+ 11
	69'11"	71'11"
Initial trim 2' aft		
COT is $\frac{165 \text{ st}}{410 \text{ st·ft/in.}} \times 10 \text{ ft} = 4"$ →	+ 2"	– 2"
New draft	70'01"	71'09"
New trim = 1'08" aft		

Example. Find the angle.

Rig displacement = 15,000 tonnes. 105 tonnes is shifted 43 m aft. KM_L = 118 m. KG = 19 m. Find the angle of inclination.

$$GG' = \frac{w \times d}{W} = \frac{105 \text{ t} \times 43 \text{ m}}{15{,}000 \text{ t}} = 0.30 \text{ m}$$

$$GM_l = KM_L - KG = 118 \text{ m} - 19 \text{ m} = 99 \text{ m}$$

$$\tan\theta = \frac{GG'}{GM_L} = \frac{0.30 \text{ m}}{99 \text{ m}} = 0.00303$$

$$\text{Angle of trim aft} = 0.17°$$

Example. Find new draft and trim.

A semisubmersible has a draft of 80 ft even keel. MT1° = 6,095 ft•st. TPI = 11.25.

Displacement = 25,070 st. VCG = 60 ft. Length = 295 ft.

75 st is loaded 82 ft aft of COF. The rig center and COF coincide with the centerline.

	Fwd	Aft
Initial draft	80'00"	80'00"
Draft increase $\frac{75\text{ st}}{11.25\text{ st/in.}}$ =	+ 6.6	+ 6.6
New average draft	80'06.6"	80'06.6"
Correction for trim as calculated below	– 2'07"	+ 2'07"
Final draft	77'11.6"	83'01.6"

Trim moment w × d = 75 st × 82 ft

$$\text{Trim} = \frac{w \times d}{MT1^\circ} = \frac{75\text{ st} \times 82\text{ ft}}{6{,}095} = 1.01^\circ$$

$$\text{To convert degrees to ft} \rightarrow \tan 1.01^\circ = \frac{\text{trim (ft)}}{L\text{ (ft)}} \quad \text{or trim} = L \tan 1.01^\circ$$

Trim = 5.2 ft

Half trim = 2.6 ft = 2'07"

The calculations for heel are similar but, of course, the moments used are for MH1°.

3.12 TANK CAPACITY TABLES AND CURVES

Tank capacity tables in Figure 3.2 and Tables 3.5 and 3.6 are examples of tank and compartment capacities. The tables and curves provide the necessary information for

1. weight, position of G for vertical, longitudinal and transverse planes with the corresponding moments.
2. permeability.
3. quick information in case of damage and/or unintentional flooding of the tank.

The figures are supplied also as information for damage stability conditions in case of any uncontrolled flooding of compartments or tanks.

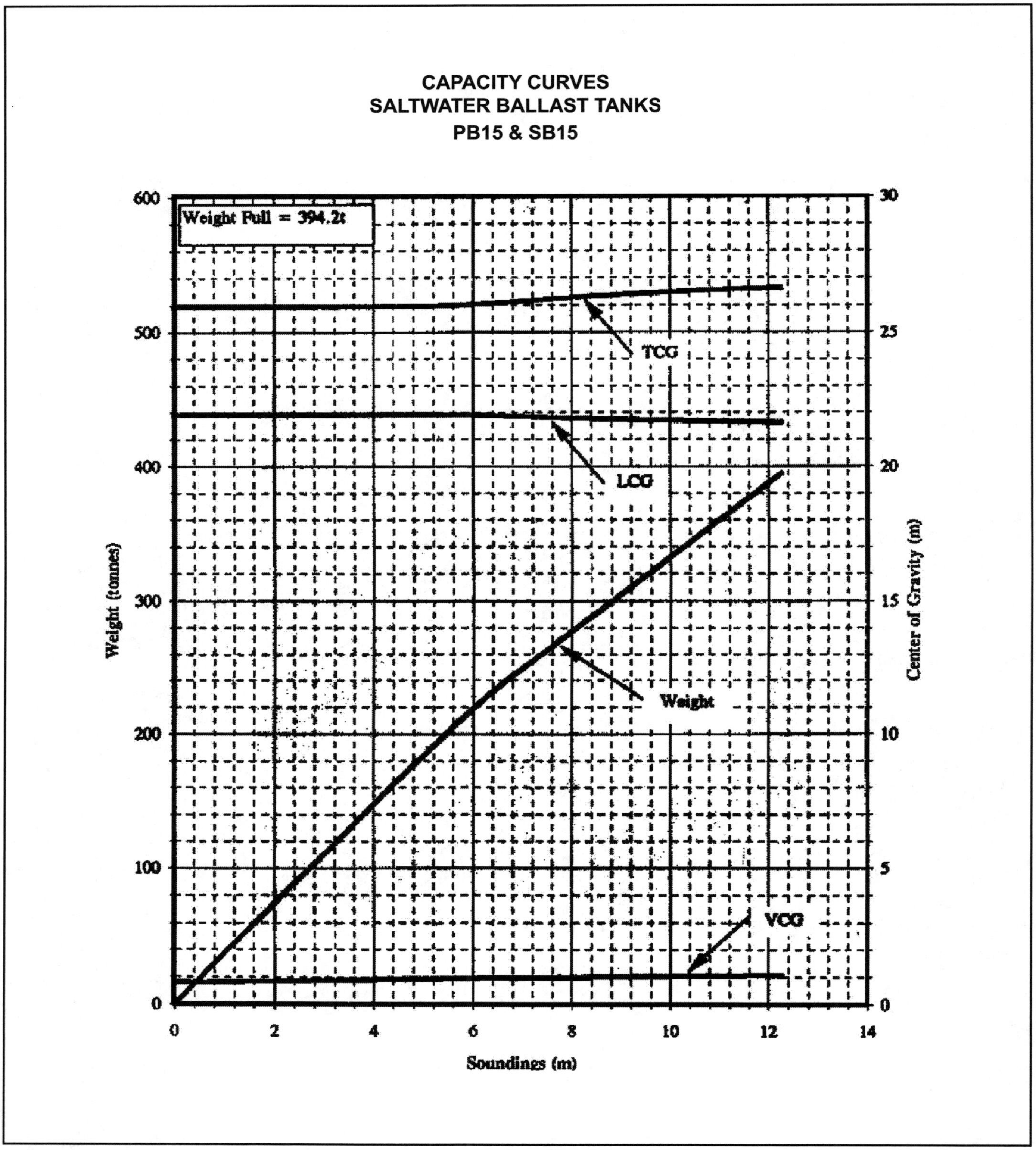

Figure 3.2 Example of a tank table curve from Rig Type 1

TABLE 3.5

Example of Capacity Table of Rig Type 1

WEIGHT AND MOMENT EFFECT OF FLOODED COMPARTMENTS (Metric)

Compartment	perm	(1)	Vertical Moment		Longitudinal Moment		Transverse Moment	
		Weight	VCG	WxVCG	LCG	WxTCG	TCG	WxTCG
	(%)	(t)	(m)	(t m)	(m)	(t m)	(m)	(t m)
Ballast Tank 1 (p&s)	98	492.8	5.19	2,558	102.01	50,271	+/- 27.50	+/- 13,552
Ballast Tank 2 (p&s)	98	668.6	5.20	3,477	93.64	62,607	+/- 30.76	+/- 20,566
Ballast Tank 3 (p&s)	98	223.1	8.64	1,928	93.44	20,846	+/- 24.22	+/- 5,403
Ballast Tank 4 (p&s)	98	345.3	2.38	822	83.50	28,836	+/- 29.75	+/- 10,274
Ballast Tank 5 (p&s)	98	487.3	4.51	2,198	72.96	35,552	+/- 28.64	+/- 13,956
Ballast Tank 6 (p&s)	98	552.5	4.50	2,486	62.18	34,353	+/- 28.07	+/- 15,508
Ballast Tank 7 (p&s)	98	383.7	2.47	948	26.60	10,205	+/- 30.63	+/- 11,751
Ballast Tank 8 (p&s)	98	543.0	5.23	2,840	16.34	8,873	+/- 30.97	+/- 16,818
Ballast Tank 9 (p&s)	98	233.1	8.62	2,009	15.39	3,587	+/- 24.09	+/- 5,615
Ballast Tank 10 (p&s)	98	545.1	5.11	2,786	7.58	4,132	+/- 29.99	+/- 16,349
Ballast Tank 11 (p&s)	98	383.6	5.14	1,972	6.03	2,313	+/- 25.32	+/- 9,713
Brine Tank 1 (p&s)	98	96.5	4.50	434	56.66	5,470	+/- 33.72	+/- 3,255
Brine Tank 2 (p&s)	98	135.0	4.50	608	47.44	6,405	+/- 33.72	+/- 4,552
Brine Tank 3 (p&s)	98	102.1	4.54	463	38.00	3,879	+/- 33.72	+/- 3,442
Base Oil Tank (p&s)	98	182.9	4.66	852	27.00	4,937	+/- 21.27	+/- 3,889
Fuel Oil Tank 1 (p&s)	98	400.5	4.02	1,610	47.44	18,998	+/- 24.95	+/- 9,992
Fuel Oil Tank 2 (p&s)	98	400.5	4.02	1,610	42.06	16,844	+/- 24.95	+/- 9,992
Drill Water Tank 1 (p&s)	98	400.5	4.02	1,610	52.81	21,149	+/- 24.94	+/- 9,988
Drill Water Tank 2 (p&s)	98	401.3	4.03	1,617	36.68	14,720	+/- 24.95	+/- 10,012
Day Tank 1 (p&s)	98	62.1	5.25	326	86.25	5,357	+/- 21.25	+/- 1,320
Day Tank 2 (p&s)	98	62.1	5.25	326	80.75	5,016	+/- 21.25	+/- 1,320
RMT 1 (p&s)	98	123.3	4.17	514	52.81	6,512	+/- 30.88	+/- 3,808
RMT 2 (p&s)	98	126.4	4.17	527	47.44	5,996	+/- 30.87	+/- 3,901
RMT 3 (p&s)	98	126.4	4.17	527	42.06	5,316	+/- 30.87	+/- 3,901
RMT 4 (p&s)	98	123.6	4.17	515	36.69	4,535	+/- 30.88	+/- 3,817
Ballast Tank 12 (p&s)	98	295.2	21.92	6,471	81.84	24,161	+/- 32.52	+/- 9,601
Ballast Tank 13 (p&s)	98	298.4	21.88	6,530	87.56	26,131	+/- 25.51	+/- 7,613
Ballast Tank 14 (p&s)	98	403.6	21.96	8,863	29.44	11,882	+/- 32.27	+/- 13,024
Ballast Tank 15 (p&s)	98	394.1	21.44	8,450	21.62	8,521	+/- 26.63	+/- 10,495
Ballast Tank 16 (p&s)	98	471.1	12.35	5,818	25.40	11,966	+/- 28.90	+/- 13,615
Potable Water Tank (p&s)	98	303.9	12.55	3,814	84.62	25,717	+/- 29.53	+/- 8,975
Electrical Room Pontoon (p&s)	85	326.7	4.50	1,470	58.00	18,950	+/- 26.75	+/- 8,740
Engine room Pontoon (p&s)	85	1143.1	4.50	5,144	69.25	79,158	+/- 26.75	+/- 30,577
Thruster Room Aft (p&s)	85	409.0	3.48	1,423	15.08	6,167	+/- 24.09	+/- 9,852
Thruster Room Fwd (p&s)	85	386.4	3.49	1,349	93.74	36,221	+/- 24.25	+/- 9,370
RMT passageway (p&s)	98	306.2	7.56	2,315	44.75	13,701	+/- 29.86	+/- 9,142
Aft Column (p&s)	98	2979.7	13.30	39,631	26.98	80,394	+/- 27.61	+/- 82,271
Fwd Column (p&s)	98	2561.6	13.87	35,529	83.53	213,968	+/- 27.55	+/- 70,571
Void space Aft Column (p&s)	98	400.5	17.23	6,900	31.86	12,759	+/- 23.64	+/- 9,467
Void space high Aft Column (p&s)	98	50.8	23.56	1,198	27.00	1,372	+/- 22.67	+/- 1,152
Void space Fwd Column (p&s)	98	348.3	17.45	6,078	80.32	27,975	+/- 23.03	+/- 8,021
Tubular Aft	98	142.0	13.50	1,917	34.00	4,829	0.00	0
Tubular Fwd	98	142.0	13.50	1,917	78.00	11,079	0.00	0
Cross Tubular (Aft Starboard to Fwd Port)	98	160.0	11.25	1,800	56.00	8,961	0.00	0
Cross Tabular (Aft port to Fwd Starboard)	98	157.9	11.25	1,776	56.00	8,843	0.00	0

NOTE:

1. The weight of liquid in tanks has been computed assuming the tanks have been damaged and therefore filled with sea water.

TABLE 3.6

Examples of Capacity Tank Tables for Rig Type 2A and Rig Type 4

Capacities of Ballast Water Tanks

Tank No.	Location in Hulls	Depth 100% Full (ft)	Tank Capacities (ft^3)	(U.S. Gal)	(ST)
PB1A	Port Bow to Fr 52 Inbd.	21.0	12332	92250	395
SB1A	Stbd Bow to Fr 52 Inbd.	21.0	12332	92250	395
PB1B	Port Bow to Fr 52 Outbd.	21.0	10960	81986	351
SB1B	Stbd Bow to Fr 52 Outbd.	21.0	10960	81986	351
PB2	Port Fr 52 to 44 Inbd.	21.0	20270	151629	649
SB2	Stbd Fr 52 to 44 Inbd.	21.0	20270	151629	649
PB3	Port Fr 52 to 44 Outbd.	21.0	20270	151629	649
SB3	Stbd Fr 52 to 44 Outbd.	21.0	20270	151629	649
PB4	Port Fr 44 to 37 Inbd. (1)	35.0	20754	155248	664
SB4	Stbd Fr 44 to 37 Inbd. (1)	35.0	20754	155248	664
PB5	Port Fr 44 to 37 Outbd.	21.0	17724	132581	567
SB5	Stbd Fr 44 to 37 Outbd.	21.0	17724	132581	567
PB6	Port Fr 37 to 29 Outbd.	21.0	20270	151629	649
SB6	Stbd Fr 37 to 29 Outbd.	21.0	20270	151629	649
PB7	Port Fr 22 to 15 Inbd. (1)	35.0	20754	155248	664
SB7	Stbd Fr 22 to 15 Inbd. (1)	35.0	20754	155248	664
PB8	Port Fr 22 to 15 Outbd.	21.0	17724	132581	567
SB8	Stbd Fr 22 to 15 Outbd.	21.0	17724	132581	567
PB9	Port Fr 15 to 7 Outbd.	21.0	20270	151629	649
SB9	Stbd Fr 15 to 7 Outbd.	21.0	20270	151629	649
PB10A	Port Fr 7 to Stern Inbd.	21.0	12332	92250	395
SB10A	Stbd Fr 7 to Stern Inbd.	21.0	12332	92250	395
PB10B	Port Fr 7 to Stern Outbd.	21.0	10960	81986	351
SB10B	Stbd Fr 7 to Stern Outbd.	21.0	10960	81986	351
PDW2	Port Fr 29 to 22 Outbd. (2)	21.0	17724	132581	567
SDW2	Stbd Fr 29 to 22 Outbd. (3)	21.0	17724	132581	567

Notes:

1. Includes 18 ft column up to elevation 35 ft.
2. Combination drill water/ballast tank, capacities shown are for salt water.
3. Combination brine/ballast tank, may be configured for drill water, capacities shown are for salt water.

2

FORAN SYSTEM - MODULE VOLUME VERSION 30 DATA

CUSTOMER - SONAT O. DR. INC
DESCRIPTION OF SHIP - D. P. DRILLING UNIT

CAPACITY TABLES

DENSITY OF THE LOAD (DN) = .920 T/M3
STEEL DISCOUNT FACTOR = .500 %

COMPARTMENT 1' CRUDE OIL N.1 FR.61-65 C.

HEIGHT M	AREA FLOT. M2	VOLUME NET M3	WEIGHT T	C. OF GRAVITY VCG M	LCG A.P. M	VOLUME TCG M	I.T.*DN M*T
.000	.00	.0	.0	.00	.00	.00	.0
1.000	475.20	472.8	435.0	3.48	176.15	.00	20984.8
2.000	475.20	945.6	870.0	3.98	176.15	.00	20984.8
3.000	475.20	1418.5	1305.0	4.48	176.15	.00	20984.8
4.000	475.20	1891.3	1740.0	4.98	176.15	.00	20984.8
5.000	475.20	2364.1	2175.0	5.48	176.15	.00	20984.8
6.000	475.20	2836.9	2610.0	5.98	176.15	.00	20984.8
7.000	475.20	3309.8	3045.0	6.48	176.15	.00	20984.8
8.000	475.20	3782.6	3480.0	6.98	176.15	.00	20984.8
9.000	475.20	4255.4	3915.0	7.48	176.15	.00	20984.8
10.000	475.20	4728.2	4350.0	7.98	176.15	.00	20984.8
11.000	475.20	5201.1	4785.0	8.48	176.15	.00	20984.8
12.000	475.20	5673.9	5220.0	8.98	176.15	.00	20984.8
13.000	475.20	6146.7	5655.0	9.48	176.15	.00	20984.8
14.000	475.20	6619.5	6090.0	9.98	176.15	.00	20984.8
15.000	475.20	7092.4	6525.0	10.48	176.15	.00	20984.8
16.000	475.20	7565.2	6960.0	10.98	176.15	.00	20984.8
16.520	.00	7811.0	7186.2	11.24	176.15	.00	.0

SIGN OF TCG: SB = - PS = +

3.13 LOADING CONDITIONS

The load-stability conditions are based on three draft conditions.

1. The operations draft.
2. The transit draft.
3. The survival draft.

3.14 THE LOAD FORM—INTRODUCTION

The load and stability conditions on a rig are constantly changing. Loading conditions and stability properties should be fully completed on a daily basis for all modes of operations, whether on DP, moored, or underway. The load form is a standard spreadsheet specific for each rig. The basic information on the load form is similar for each rig.

3.15 LOAD FORM PURPOSE AND PROCEDURE

The purpose of the load form is to enter the details of the variable load in tabulated pages for tanks including free surface effect, drilling loads, deck loads, bulk storage, stores, and, if applicable, the anchor system corrections.

The basic stability calculation is similar for all stability calculations:

1. Find for each line the weight, VCG, LCG, and TCG.
2. To find the moments, multiply the weight with the VCG, LCG, and TCG.
3. Add for tanks with liquid the free surface moments.
4. Enter at the bottom of each page the total weight, VCG, LCG, TCG, and corresponding moments.
5. Enter the totals of each page in the summary section of page one.
6. The final result is:

$$VCG = \frac{\text{Final Vertical Moments}}{\text{Final Displacement}}$$

$$LCG = \frac{\text{Final Longitudinal Moments}}{\text{Final Displacement}}$$

$$TCG = \frac{\text{Final Transverse Moments}}{\text{Final Displacement}}$$

3.16 LOAD FORM SUMMARY PAGE

A final page with lightship weight, and the total results of each individual page summarizes the total displacement (deadweight) and corresponding total moments (VM) with the position VCG, LCG, and TCG including the VGG_{wet} with the free surface.

3.17 MAXIMUM ALLOWABLE VCG

In addition to the information on the final summary page, the VGG_{wet} is compared with the maximum allowable VCG. The maximum allowable VCG is the result of the stability analysis as a safety margin to ensure the rig under all conditions has enough reserve stability. The maximum allowable VCG value for a specific draft and condition ensures enough reserve stability is available for operations and survival conditions including the damage stability condition. The maximum allowable VCG value will be explained in detail later on. Figure 3.3 is a typical example of the summary page from the Rig Type 4.

Before computers entered the offshore industry, the load form was a hard copy to enter manually all weights and gravity centers. Computers on board now perform all the calculation on mimics of the hard copy load form. Data grouping allows much faster calculation and provides immediate computing of the final results. Once a person understands how to enter load data in the computer, he virtually can complete the stability information without understanding the basic principles of stability. It is most important that the daily stability calculations are done accurately by a qualified person in charge of and responsible for the stability calculations. A good practice is to verify at regular intervals the data with a manual calculation on the hard copy spreadsheets.

DEADWEIGHT SUMMARY									
No	Item	Weight	VCG	VM	LCG	LM	TCG	TM	FSM
8	Cargo Tanks	0	0	0	0	0	0	0	213
28	Ballast Tanks	45,932.0	8.15	374,250	120.87	5,551,963	0.13	5,777	2,068
38	Diesel Oil Tanks	3,992.6	9.67	38,611	48.40	193,260	0.11	451	3,838
50	Fresh Water & Misc Tanks	2,218.9	8.17	18,123	87.80	194,809	0.25	565	1,859
213				0		0		0	
214	Subtotal, Hull Loads	52,143.5	8.27	430,984	113.92	5,940,032	0.13	6,793	7,978
69	Res Dr Fluids & Bulk Mat'l	1,157.3	7.08	8,194	91.16	105,505	-0.31	-357	63
103	Active & Ret Drill Fluids	913.4	24.08	21,995	85.74	78,311	-0.17	-157	211
157	Riser, Pipe Rack Loads	323.0	26.84	8,670	98.35	31,766	0.01	4	
183	Drilling Loads	2,276.1	44.24	100,692	132.69	302,009	-2.30	-5,226	
212	Stores & 3rd Party Loads	2,167.7	29.23	63,370	106.73	231,366	-2.76	-5,973	
215	Snow & Ice Loads			0		0		0	
216				0		0		0	
217	Subtotal, Deck Load	6,837.5	29.68	202,921	109.54	748,958	-1.71	-11,709	274
218	**TOTAL, DEADWEIGHT**	**58,981.1**	**10.75**	**633,905**	**113.41**	**6,688,989**	**-0.08**	**-4,916**	**8,252**

Table 3-2 – GOM Exploration Well, Operating Condition (Continued)

VESSEL LOADING SUMMARY									
No	Item	Weight	VCG	VM	LCG	LM	TCG	TM	FSM
219	Lightship Weight 1999 IE	34,135.9	20.49	699,275	117.41	4,007,935	0.20	6,861	
220	Lightship Amendment			0		0		0	
218	Total, Deadweight	58,981.1	10.75	633,905	113.41	6,688,989	-0.08	-4,916	8,252
221				0		0		0	
222	**DISPLACEMENT**	**93,117.0**	**14.32**	**1,333,180**	**114.88**	**10,696,924**	**0.02**	**1,945**	**8,252**

Coordinates and References:

VCG from BL, + up
LCG from AP (Fr. No. 0), + forward
TCG from CL, + port
Units: metric tons, m

Stability Reduction:

Displacement	93,117.00	MT	From Vessel Loading Summary, Line 222
Draft @ LCF	12.00	m BL	From Hydrostatic Data, Appendix B
VCG (or KGo)	14.32	m BL	From Vessel Loading Summary
KMT	16.51	m BL	Transverse Metacentric Height, from Hydrostatic Data
GMT	2.19	m	GMT = KMT – KGo
Free Surface Corr, FSC	0.09	m	FSC = FSM/Displ, FSM from Vessel Loading Summary
GMT Corr	2.10	m	GMT Corr = GMT – FSC
KG Corr	14.41	m BL	KG Corr = KGo + FSC
KG Allowable	15.44	m BL	From Allowable KG Curve, Section 2.4
Margin on KG Allow	1.03	m	KG Allowable – KG Corr

Trim & Drafts Reduction:

LCG	114.88	m AP	From Vessel Loading Summary, Line 222
LCB	114.69	m AP	From Hydrostatic Data, Appendix B
Trim Lever	-0.19	m	LCB - LCG
MT1cm	1,631.10	MT-m	Moment to Trim 1 cm, from Hydrostatic Data
Trim (+ aft, - fwd)	-0.11	m	Trim = (Displ x Trim Lever)/(100 x MT1cm)
LCF	115.67	m AP	From Hydrostatic Data
Draft @ AP	11.95	m BL	Draft @ AP = Draft @ LCF + (Trim x LCF/240)
Draft @ FP	12.06	m BL	Draft @ FP = Draft @ AP – Trim
Draft @ Midships	12.00	m BL	Draft @ Midships = (Draft @ AP + Draft @ FP)/2

Figure 3.3 Example of Load form summary page Rig Type 4

3.18 TOTAL DISPLACEMENT AND DRAFT

The total displacement must correspond within limits with the actual draft readings. Some rig's marine operation manuals give instructions as to what action is required if the difference between the actual and calculated values is beyond the acceptable limits.

As an example for Rig Type 2, the calculated displacement must be within 200 st of the displacement corresponding with the actual mean draft displacement. With respect to trim and heel, the LCG and TCG values must be within one foot of the actual value.

In general, only minor discrepancies are acceptable. If the difference exceeds the allowable tolerance, the following action is required:

1. Recheck the arithmetic of the calculation. If necessary, perform a manual calculation.
2. Recheck all input from the computer.
3. Recheck all deck loads and their positions.
4. Take manual soundings of all tanks and bulk storage.
5. If there is still a major difference, check all void spaces.
6. If the reason for the discrepancy remains unsolved, forward a copy of the calculation to the engineering department (naval architect) for further advice.

3.19 THE MAXIMUM ALLOWABLE AREA LOADING

The strength of the deck at various levels is not the same. The main deck generally has the highest strength, as this deck is particularly used to store drilling equipment, tubulars, and other stores.

To prevent damage to the deck and girders, heavy load concentration should be avoided. The load concentration can be spread with the use of supporting timbers under the loads. Figures 3.4 and 3.5 are examples of deck strength indicating the maximum area loading on Rig Type 2.

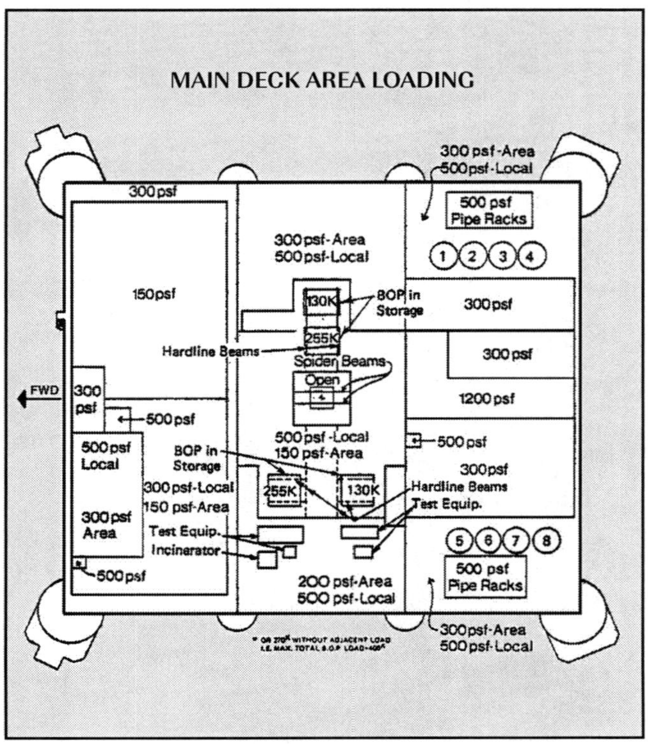

Figure 3.4 Rig Type 2—Maximum allowable area loading Imperial system

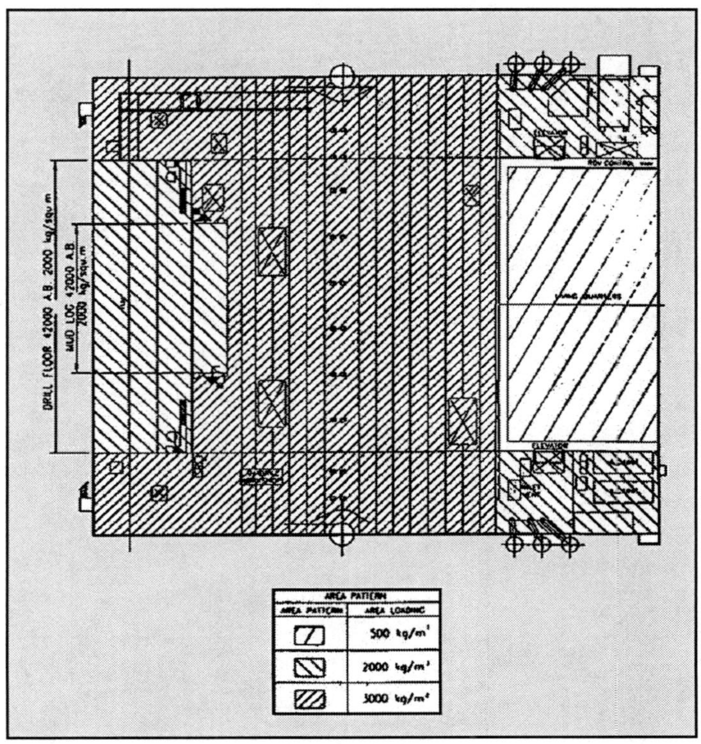

Figure 3.5 Rig Type 2—Maximum allowable area loading metric system

EXERCISE 1
Chapter 3

1. A rig with a displacement of 17,950 mt and an even keel draft of 19.10 m loads 410 mt. TPC = 34.5 t. Find the new draft.

2. A vessel floats at a draft of 25'04" fwd and 27'02" aft. MTI" = 7,190 ft•st. LCF is amidships. A weight of 410 st is shifted forward over a distance of 245 ft. Find the COT and the new drafts.

3. A drilling vessel floats at 8.30 m fwd and 8.80 m aft. The length is 116 m. TPC = 18. MCT 1 cm = 112 t•m/cm. The LCF is 2.2 m aft of amidships. Find the new drafts after the following weights have been loaded or discharged.
   a. Loaded 60 mt at 13 m fwd of the centerline
   b. Loaded 210 mt at 18 m from the stern
   c. Discharged 55 mt 18 m aft of the centerline

Use Table 3.2 to answer the following questions.

4. Find the increase in displacement if the draft changes from 20.04 m to 20.90 m.

5. How many tonnes need to be discharged to change the draft from 20.60 m to 19.74 m?

6. At a displacement of 34,850 t, a weight of 834 t is discharged. What is the average new draft?

7. At a draft of 19.85 m, what is:
   a. Displacement
   b. VCB
   c. LCB
   d. TCB
   e. WPA
   f. TPC
   g. LCF
   h. TCF
   i. KM_L
   j. KM_T
   k. MCT
   l. MCH

EXERCISE 2
Chapter 3

Use the hydrostatic Table 3.1 Rig Type 2A at page 3.2 to answer the following questions.

Assume: Width between the draft marks = 195 ft
Length between the draft marks = 225 ft

1. With a draft of 70 ft upright (no heel and trim) and a KG of 60 ft, a load of 40 st on the port deck is moved from a position 95.5 ft off the centerline and 130 ft above the keel to the same offset position to starboard. Calculate the final angle of list.

2. Rig Type 2A is at 72 ft draft upright, KG = 61.95 ft. The rig loads 200 st on the port deck, 47 ft off centerline and 130 ft above the keel. Find the new average mean draft, and the angle of list.

3. Rig Type 2A is at 65 ft draft with a list of 2.5° to port. KG = 65 ft. Find the distance that 25 st of cargo needs to be moved transversely from the port main deck to the starboard main deck to get zero list. Is it safe to do this?

4. The rig is at 70.5 ft draft with zero heel and trim. KG = 64.16 ft. A total weight of 335 st is loaded at 27.5 ft starboard of the longitudinal center. The VCG of the load is 130 ft above the keel. Given the TCF is on the centerline of the rig, calculate the final draft at each corner.

5. Rig Type 2A floats at a draft of 76 ft with zero degree heel and trim. KG = 65 ft. A load of 80 st is moved 40 ft aft and 30 ft to starboard. Determine the final draft at each corner.

6. Rig Type 2A floats at a draft of 72.5 ft with zero heel and trim. KG = 60.06 ft. A total load of 267 st is placed at 95 ft to starboard of the TCF and 100 ft forward of the LCF. The VCG of the load is 54.08 ft above the keel. Find the
   a. new KG.
   b. sinkage.
   c. new displacement and corresponding mean draft.
   d. angle of heel.
   e. change of heel in ft.
   f. angle of trim.
   g. change of trim in ft.
   h. draft on each column corner.

7. **Do this question only if required.** Rig Type 2A is floating at a draft of 70 ft with zero degree heel and trim. KG = 61.2 ft. The rig loads 135 st at 75 ft aft of the centerline and 96 ft to starboard of the centerline. The VCG of the load is 132 ft above the keel. Calculate the final draft on each corner.

CHAPTER 4
The Mooring System

4.1 INTRODUCTION

The main purpose of the mooring system is to maintain the rig position over the well within the operating limits of the BOP and riser system. For all conditions, the rig will move off the zero position until the restoring forces of the mooring system are equal to the energy applied by the environmental forces.

4.2 OPERATION CONDITIONS

Under normal operation conditions, the restoring forces from the mooring system maintain the rig position within the operating limits of the drilling equipment.

4.3 MAXIMUM OPERATING CONDITIONS

The maximum operating condition is restricted to the

1. maximum allowable angle between the marine riser system and the BOP.
2. maximum stroke of the heave compensating system.
3. capacity of the riser tensioning system.
4. pretest tension.

The maximum allowable offset is, in most cases, expressed in percentage of water depth. The mooring analysis includes the anchor tension corresponding with the maximum allowable offset.

4.4 SURVIVAL CONDITION

In addition to the operation conditions, the mooring system capacity must be able to maintain the rig safely moored in severe weather and current conditions with the riser system disconnected from the seabed.

4.5 THE MOORING SYSTEM LIMITS

The mooring system itself is limited by the

1. holding power of the anchor.
2. length of the deployed chain.
3. safe cable tension.

4.6 THE ANCHOR SYSTEM TENSIONS

The anchor system tension values and corresponding restoring forces are based on the

1. survival condition tension.
2. maximum operating tension.
3. optimum operating tension.

4.7 THE SURVIVAL CONDITION TENSION (fig. 4.1)

The maximum tension on the mooring system should not exceed 50% of the break load test. This is called the survival tension. For 3" ORQ anchor chain, this value is rounded off to 500,000 lb. At the survival tension value, the chain or wire is stretched to a straight line towards the anchor. The anchor shank starts to lift and the anchor will tend to slip instead of holding up the system to tension higher than 500,000 lb. The 100 kt wind is used as the criteria for the survival tension.

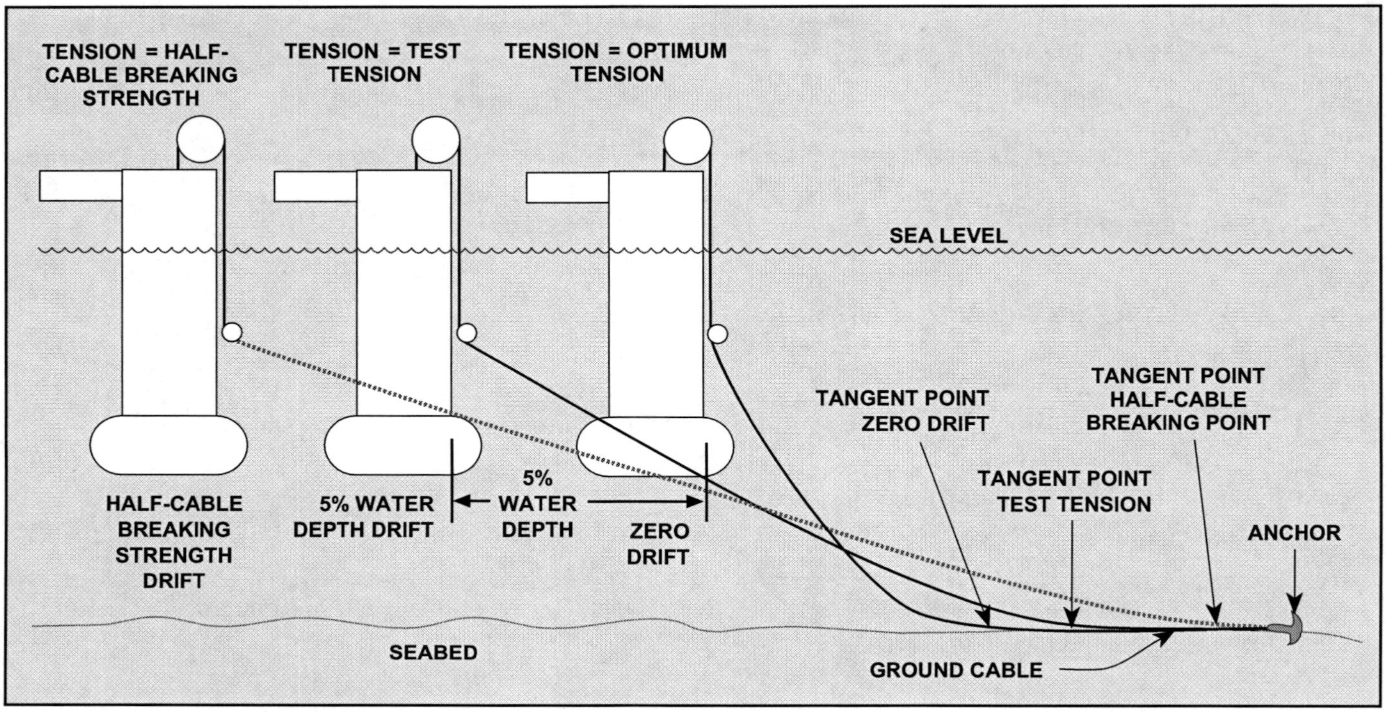

Figure 4.1 Survival condition tension

4.8 THE MAXIMUM OPERATING TENSION (fig. 4.2)

The proof load of the anchor chain is 66% of the breaking strength. For 3" ORQ chain, the proof load is rounded off to 700,000 lb. The maximum operation tension should not exceed 50% of the proof load test. This is referred to as test or pretest tension applied to the anchor system before the start of the drilling operation. For 3" ORQ chain, this value is rounded off to 350,000 lb. For 3" wire, this value is rounded off to 240,000 lb.

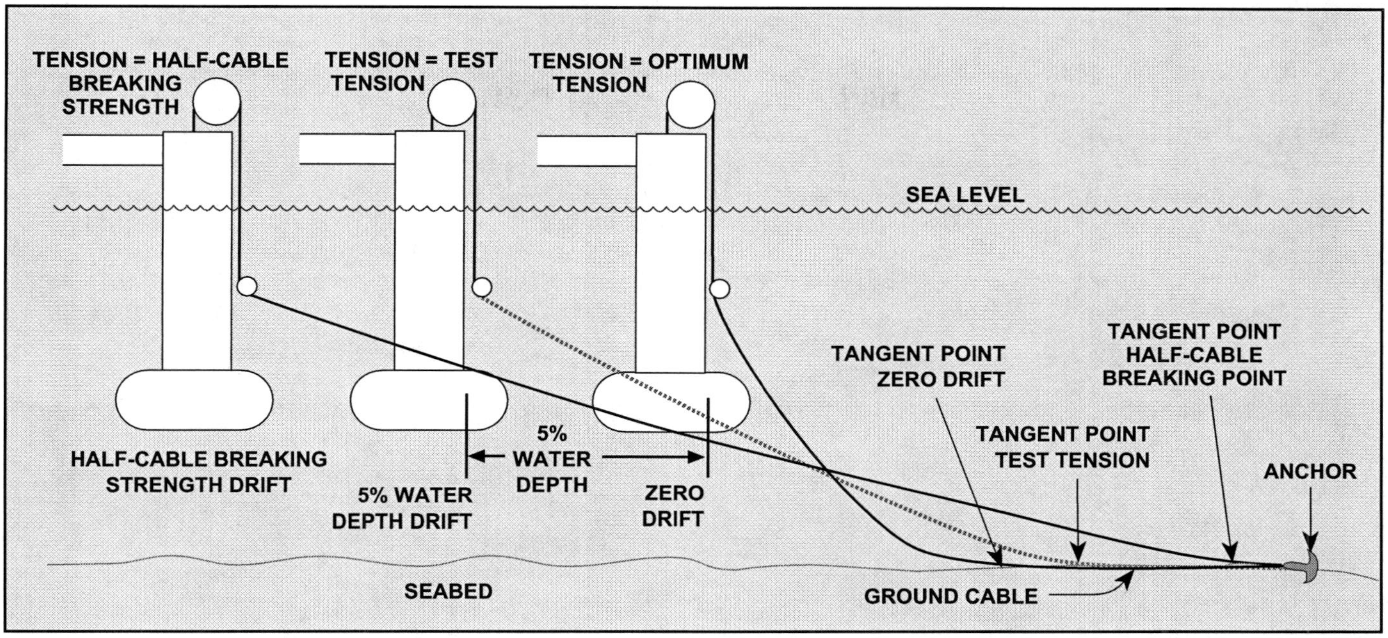

Figure 4.2 The maximum operating tension

4.9 THE OPTIMUM TENSION (fig. 4.3)

The maximum allowable offset (for the BOP angle, riser tension, heave compensatory) should coincide with the maximum operations tension (350,000 lb for 3" chain).

After successful completion of the test tension (350,000 lb for 3" anchor chain or 240,000 lb for 3" wire), the optimum tension is applied to all anchors. With the optimum tension, the anchor line tension will reach the test tension value when the rig offset equals the maximum allowable offset for the riser system or maximum angle for the BOP system.

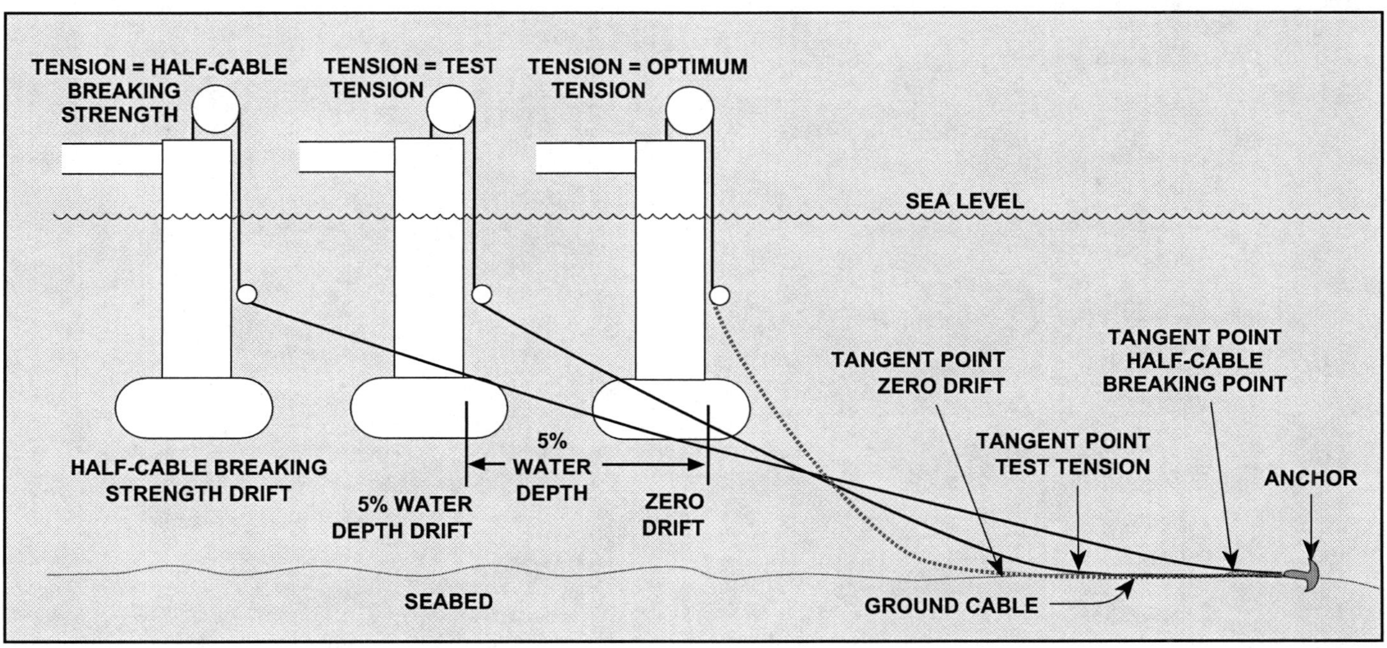

Figure 4.3 The optimum tension

4.10 TABLES WITH CHAIN LENGTH FOR SURVIVAL TENSION AND OPTIMUM TENSION

The rig's marine operations manual tables display the minimum length of chain to run out versus water depth for the maximum survival tension and separate tables for the optimum tensions versus water depth for various percentages of offset (Table 4.1 and Table 4.2).

4.11 THE HOLDING POWER OF THE MOORING SYSTEM

The total holding power of the mooring system is the sum of the

1. anchor-holding power.
2. holding power of the chain or wire.

$$\text{The efficiency of the anchor} = \frac{\text{The maximum holding power before the anchor starts to slip}}{\text{Weight of the anchor}}$$

4.12 THE ANCHOR-HOLDING POWER

The efficiency of an old type of anchor like the LWT anchor is about 12 compared to the modern Vryhof and Bruce anchors with an efficiency of more than 40.

4.13 THE HOLDING POWER OF THE CHAIN OR WIRE

The holding power of the chain or wire depends on the

1. length of the anchor cable on the seabed.
2. weight of the anchor cable.
3. friction coefficient of the anchor cable.
4. soil condition.

In formula, the approximate holding power of the anchor cable is:

$$P_c = f \cdot L_c \cdot W_c$$

P_c = holding power

f = friction coefficient

L_c = length of the anchor cable on the seabed

W_c = weight of the chain in seawater

For example W_c for 3" chain is 78.2 lb/ft and 3" wire 15.7 lb/ft

Table 4.3 displays the friction coefficients for chain and wire.

4.14 MOORING PATTERNS

To maintain the rig's position within the operations and survival criteria, the symmetrical anchor pattern provides for the best flexible mooring system to absorb the environmental forces from any direction. Figure 4.4 shows some typical symmetrical anchor patterns.

TABLE 4.1

Minimum Length of Chain Versus Water Depth for Rig Type 2A

RIG TYPE 2A

DISTANCE FROM UNIT CENTER TO ANCHOR PATTERN BUOY (D)
AND MINIMUM LENGTH OF CHAIN RUNOUT (L)
VS
WATER DEPTH

Water Depth at Anchor (ft)	D (ft)	L (ft)
200	2,480	1,500
300	2,860	1,880
400	3,160	2,200
500	3,410	2,470
600	3,630	2,710
700	3,840	2,930
800	4,000	3,130
900	4,160	3,310
1,000	4,300	3,480
1,200	4,580	3,780
1,400	4,840	4,060
1,600	5,080	4,310
1,800	5,320	4,560

L is the length of chain runout that produces a chain tension equal to 500,000 lb (50% of breaking strength) when all chain just lifts off the seafloor. D is approximately 1,000 ft plus the horizontal distance corresponding to L.

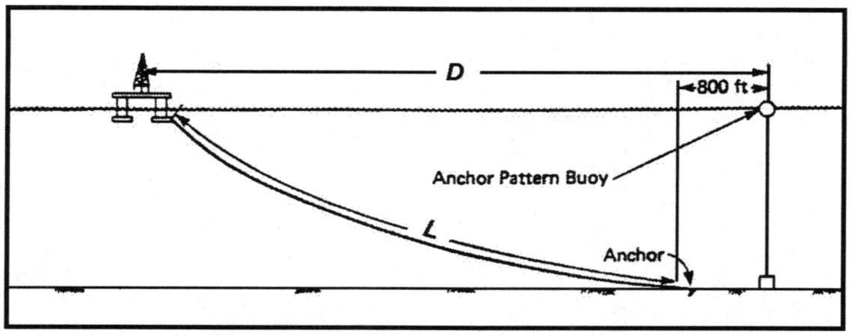

TABLE 4.2

Optimum Tension Values Versus Water Depth Rig Type 2A

The following is a typical example of optimum tension values versus offset limits from Rig Type 2A's marine operations manual:

Water depth (ft)	Test Tension (lb)	Appr. Optimum Tension (lb) 3% WD	5% WD	7% WD
200	350,000	225,000	170,000	130,000
400	"	250,000	210,000	175,000
600	"	270,000	230,000	200,000
1000	"	290,000	260,000	235,000
1200	"	295,000	270,000	245,000
1600	"	305,000	285,000	255,000

TABLE 4.3
Friction Coefficients (f) for Anchor Chain and Wire

Ocean Bottom	Friction Coefficients Starting	Sliding
Chain Holding Power		
Sand	0.98	0.74
Mud with Sand	0.92	0.69
Firm Mud	1.01	0.62
Soft Mud	0.90	0.56
Clay	1.25	0.81
Wire Rope Holding Power		
Sand	0.98	0.25
Mud with sand	0.69	0.23
Firm Mud	0.50	0.21
Soft Mud	0.45	0.18
Clay	0.63	0.27

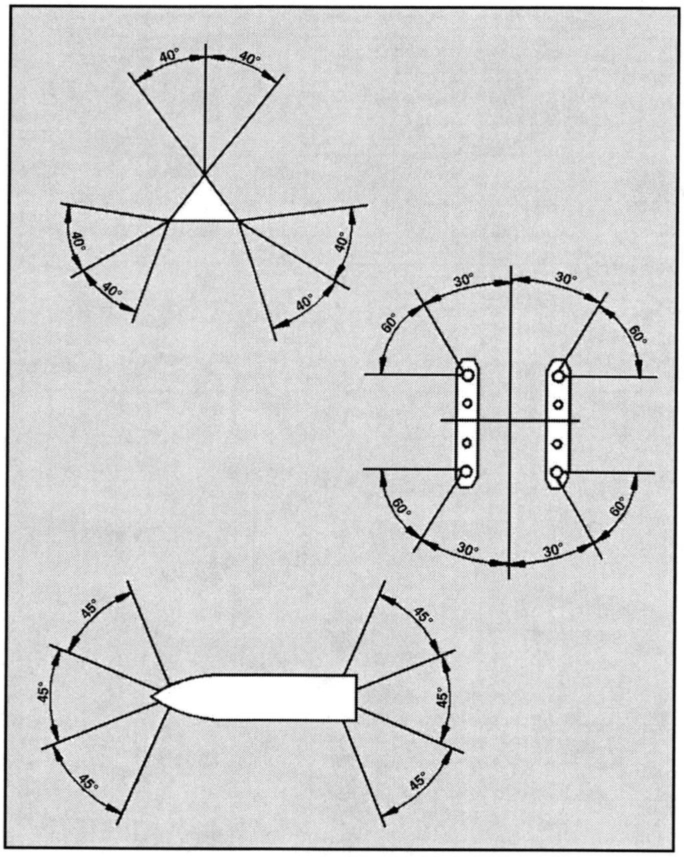

Figure 4.4 Symmetrical anchor patterns

4.15 THE CATENARY SYSTEM (fig. 4.5)

The form of perfectly flexible cord of uniform density and shape suspended freely between two fixed points is called a catenary curve. The curve of the anchor cable between the fairleader and the touchdown point at the seabed is half a catenary curve.

The shape depends on the

1. horizontal distance between the fairleader and touchdown point.
2. length of the anchor cable.
3. weight of the anchor cable.

Other important information is:

- The distances, tensions, and angles can be calculated.
- The rig maintains position by opposite horizontal components of the tension.
- The *vertical component* (*V*) of the tension is the weight of the chain. *V* reduces on the way down.
- The catenary system works as a spring.
- The anchor analysis is a computerized calculation of the anchor system behavior.

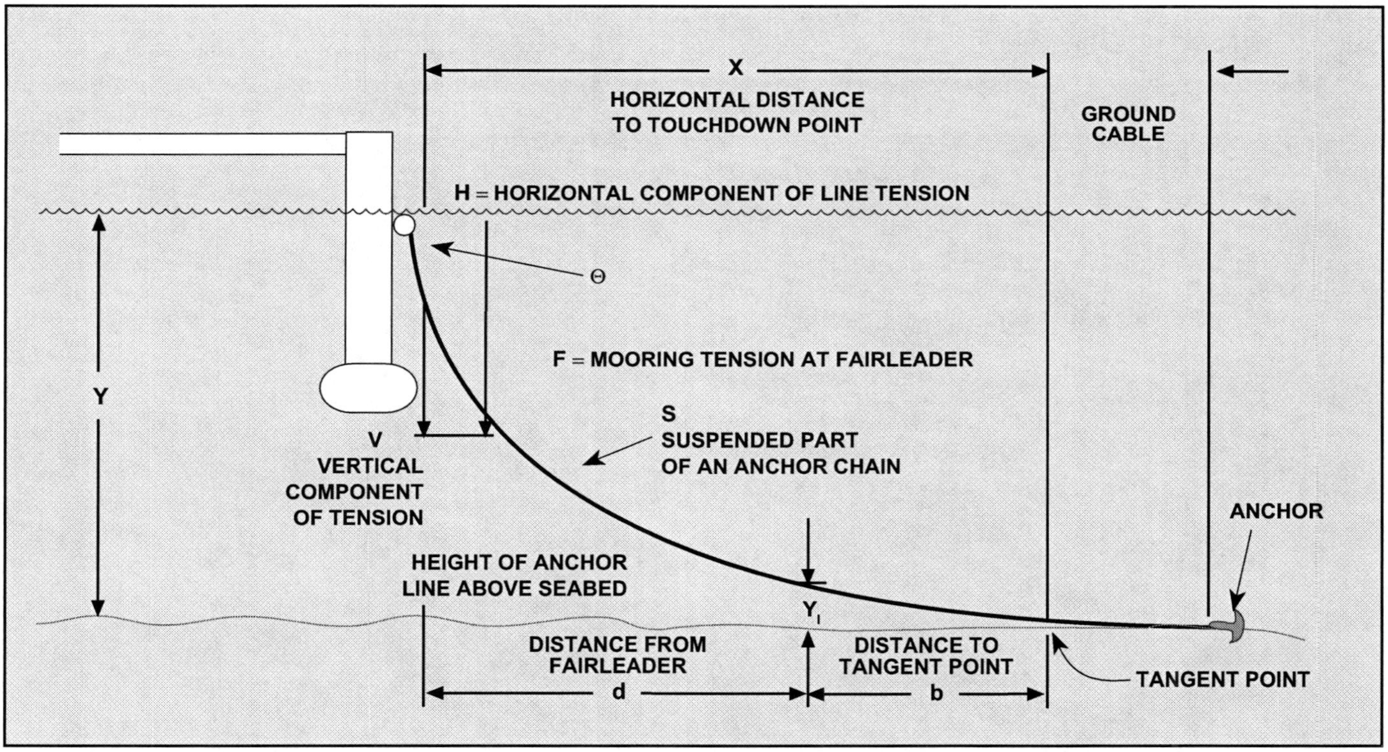

Figure 4.5 The catenary system

4.16 THE WEIGHT OF THE SUBMERGED ANCHOR CABLE

The submerged weight of the anchor line can be calculated.

Example of calculation of submerged weight of 3" chain:
The density of steel is approximately 490 lb/ft^3.
The weight of 3" ORQ chain is 89 lb/ft in air.
The volume of steel per ft is $\frac{89\ lb/ft}{490\ lb/ft^3}$.
Volume per ft = 0.183 ft^3.
1 ft^3 of seawater weighs 64 lb.
1 ft of chain displaces 0.182 $ft^3 \cdot 64\ lb/ft^3$ = 11.64 lb.
The weight of the submerged 3" chain is therefore 89 lb – 11.64 lb = 77.36 lb/ft.
Another method is to use the buoyancy fact of steel which is 0.869.
Multiplying the weight of chain in air by 0.869 gives the submerged weight of the chain.

4.17 FORMULAS FOR THE CATENARY SYSTEM

The values along the catenary system can be calculated with the formulas shown below. In practice, all calculations are done with a computerized program with the various scenarios required for an anchor analysis. The anchor analysis is specific for the rig and the location. Part of the analysis is the broken line condition, which shows the new position of a rig after one of

Formulas used for the catenary system.

1. $F^2 = V^2 + H^2$
2. $\left[Y + \frac{H}{W}\right]^2 = S^2 + \left[\frac{H}{W}\right]^2$
3. $Y = \left[\frac{H}{W}\right]\left\{\cos\theta\left(\frac{W \cdot H}{H}\right) - 1\right\}$
4. $S = \left[\frac{H}{W}\right]\left\{\sin\theta\left(\frac{W \cdot H}{H}\right) - 1\right\}$
5. $V = W \cdot S$ or $V = \sqrt{2F \cdot Y \cdot W - (Y \cdot W)^2}$
6. $S = \sqrt{\frac{2F \cdot Y}{W} - Y^2}$
7. $F = H + (W \cdot Y)$
8. $X = \frac{H}{W}\, l_n \left[\frac{\left(\frac{F}{W} - S\right)}{\frac{H}{W}}\right]$

As per Figure 4.5:

F = Mooring tension at the fairleader in lb.
V = Vertical component of the tension at the fairleader in lb.
H = Horizontal component of the tension at the fairleader in lb.
Y = Vertical distance fairleader to seabed or height of the cable in ft.
W = Submerged weight of the anchor cable in lb/ft.
S = Suspended length of the anchor cable in ft.
X = Horizontal distance fairleader to touchdown point of the anchor cable.
l_n = Natural log.
θ = Angle between anchor cable and horizontal.

the anchor lines, for whatever reason, breaks. This information is supplied by the engineering department and is often part of the requirements by the client.

Although some of the formulas to calculate the catenary system are reasonably complicated, the calculations can be done on board with a scientific calculator.

4.18 THE ANCHOR SYSTEM CORRECTION

Because the lightship weight and the corresponding position of G includes the weight of chain in the chain locker and the anchors in racked position, a correction is required to compensate for the

1. amount of anchor chain and anchors deployed. This is a weight discharged. The corresponding moment will be deducted from the stability calculation.
2. vertical component of the anchor tension represents a weight at the fairleader. This weight and moment will be added to the stability calculation.

The calculations for the anchor system correction are relatively simple. The naval architect prepares Anchor System Correction Sheets and tables to apply the correction as part of the stability calculations. Although there are differences between one rig and another, the principles are the same. An example is explained below for 4,300 ft of chain on board and anchor #2 with 3,200 ft runout of 3" chain. The tension is 180 kips at the windlass for a water depth of 650 ft for the Rig Type 2A. Use Tables 4.4 and 4.5 and Figure 4.6.

TABLE 4.4

Anchor Chain Versus Moment

RIG TYPE 2A

VERTICAL MOMENT OF DEPLOYED ANCHOR CHAIN

Vertical Moment (ft•st) Chain Lockers No. 1, 4, 5, 8

		Anchor Chain on Board (ft)								
		2500	3000	3500	4000	4300	4500	4800	5000	5300
A	1000	2108	2461	2818	3175	3389	3532	3746	3889	4103
N	1200	2455	2867	3296	3724	3981	4153	4410	4581	4830
C	1400	2780	3246	3745	4245	4545	4745	5045	5245	5533
H	1600	3084	3604	4166	4737	5080	5309	5651	5880	6207
O	1800	3367	3940	4558	5201	5586	5844	6229	6486	6852
R	2000	3628	4255	4926	5636	6064	6350	6779	7064	7469
	2200	3867	4548	5273	6042	6514	6828	7299	7614	8056
C	2400		4819	5598	6421	6934	7277	7791	8134	8614
H	2600		5070	5902	6779	7327	7698	8255	8626	9143
A	2800			6184	7115	7695	8090	8690	9090	9643
I	3000			6446	7430	8041	8458	9096	9525	10114
N	3200 →			6685	7723	8367	8805	9476	9931	10556
	3400				7994	8671	9130	9833	10311	10969
D	3600				8245	8953	9434	10169	10668	11354
E	3800					9214	9717	10484	11004	11716
P	4000					9454	9978	10777	11319	12056
L	4200						10217	11049	11612	12374
O	4400							11299	11884	12671
Y	4600								12134	12946
E	4800									13199
D	5000									13431

If weight of 3" chain in air = 90 lb/ft

$$\text{Weight of chain (WT) deployed in st} = \frac{\text{Chain runout in ft} \cdot \text{weight in air}}{2{,}000}$$

$$\text{WT} = \frac{90\ \text{t}}{2{,}000}\ \text{runout ft} \rightarrow \text{WT} = 0.045 \cdot \text{runout in ft.}$$

For the example, the value is 3,200 · 0.045 = 144 st. This is a weight deduction.

The calculation for vertical component is prepared as shown in Table 4.5.

For the example, the value is 61.3 st. This is a weight addition.

The difference between the two values is the weight correction.

For the example, this is 144 - 61.3 = 82.7 st.

The corresponding moment for 144 st chain deployed is prepared in Table 4.4.

The naval architect calculated the value of Table 4.5 based on a 6 ft diameter chain locker.

This value is specific for Rig Type 2A based on the formula.

Change of moment is (H - 0.0079 · chain runout) · short tons deployed. See Figure 4.6.

H is the value in our example for the height with 4,300 ft of chain on board.

In the example, the moment for chain deployment is 8,367 st·ft.

The value for the moment for the vertical component of the tension is based on a height of the fairleader of 48 ft above keel for the Rig Type 2A. See Figure 4.6.

In the example, the moment for the vertical component of the tension is 61.3 · 48 = 2,942 st·ft.

The LM and TM are calculated with the respective LCG and TCG values.

For one chain deployment, the total correction is 82.7 st – 5,425 st·ft.

Example of anchor system correction

PAGE 13A—ANCHOR CHAIN CORRECTION

CHAIN No.	(1) CHAIN OUT (FT.)	(2) WT. OF CHAIN DEPLOYED .045 X (1) (ST)	TENSION (1000 LB.)	WATER DEPTH (FT.)	(3) TABLE 3.1 VALUE (ST)	(4) CHAIN CORRECTION (2) - (3) (ST)	(5) TABLE 3.2 VALUE (ST-FT)	5A VM 5-[(3) X 48] (ST-FT)	(6) LCG (FT.)	LM (4) X (6) (ST-FT)	(7) TCG (FT.)	TM (4) X (7) (ST-FT)
1								– =	112.5		106.5	
2								– =	121.5		97.5	
3								– =	121.5		-97.5	
4	3200	144	180	650	61.3	82.7	8367	8367 - 2942 = 5425	112.5	10048	-106.5	8022
5								– =	-112.5		-106.5	
6								– =	-121.5		-97.5	
7								– =	-121.5		97.5	
8								– =	-112.5		106.5	
						82.7		5425		10048		8022

NOTES:

1. FOR "H" VALUES SEE NOTE 3 IN *LIGHTSHIP ADJUSTMENT FOR CHAIN ADDITION/SUBTRACTION* PART

TABLE 4.5
Vertical Component of the Anchor Chain Tension, Rig Type 2A

RIG TYPE 2A
WEIGHT OF ANCHOR CHAIN SUSPENDED FROM UNIT
(FOR ONE CHAIN)

WEIGHT OF ANCHOR CHAIN (st)

Water Depth (ft)	Average Anchor Tension at Windlass (Kips) 60	80	100	120	140	160	180	200	220	240	260	280	300
150	15.8	18.4	20.7	22.8	24.7	26.5	28.1	29.7	31.1	32.6	33.9	35.2	36.5
200	18.6	21.8	24.6	27.1	29.4	31.5	33.5	35.3	37.1	38.8	40.5	42.0	43.5
250	20.8	24.5	27.8	30.6	33.3	35.7	38.0	40.1	42.2	44.1	46.0	47.8	49.5
300	22.7	26.9	30.5	33.7	36.7	39.4	41.9	44.3	46.6	48.8	50.9	52.9	54.8
350	24.2	28.9	32.9	36.4	39.7	42.7	45.5	48.1	50.6	53.0	55.2	57.4	59.5
400	25.6	30.6	35.0	38.9	42.4	45.6	48.6	51.5	54.2	56.8	59.2	61.6	63.9
450	26.7	32.2	36.9	41.0	44.8	48.3	51.6	54.6	57.5	60.3	62.9	65.4	67.9
500	27.6	33.5	38.6	43.0	47.1	50.8	54.3	57.5	60.6	63.5	66.3	69.0	71.6
550	28.3	34.7	40.1	44.9	49.1	53.1	56.8	60.2	63.5	66.6	69.5	72.4	75.1
600	28.9	35.8	41.5	46.5	51.0	55.2	59.1	62.7	66.1	69.4	72.5	75.5	78.4
650	29.4	36.7	42.7	48.0	52.8	57.2	61.3	65.1	68.7	72.1	75.3	78.5	81.5
700	29.7	37.5	43.9	49.4	54.5	59.0	63.3	67.3	71.0	74.6	78.0	81.3	84.4
750	29.9	38.1	44.9	50.7	56.0	60.8	65.2	69.4	73.3	77.0	80.6	84.0	87.2
800	30.0	38.7	45.8	51.9	57.4	62.4	67.0	71.3	75.4	79.3	83.0	86.5	89.9
850		39.2	46.6	53.0	58.7	63.9	68.7	73.2	77.4	81.5	85.3	88.9	92.5
900		39.5	47.3	54.0	59.9	65.3	70.3	75.0	79.4	83.5	87.5	91.3	94.9
950		39.8	47.9	54.9	61.1	66.7	71.8	76.7	81.2	85.5	89.6	93.5	97.3
1000		39.9	48.5	55.7	62.1	67.9	73.3	78.3	82.9	87.4	91.6	95.6	99.5

In addition, the correction for the weight and moments for anchor #2 need to be deducted. See Table 4.6.

TABLE 4.6
Correction for the Anchor Deployment

PAGE 138—ANCHORS AND SWIVELS

ANCHOR NO.	WT. (st)	VCO (ft)	VM (st-ft)	LCG (ft)	LM (st-ft)	TCG (ft)	TM (st-ft)
1	23	25	575	+112.5	+2,588	+131.5	+3,825
2	23	25	575	+132.5	+3,848	+125.5	+2,887
3	23	25	575	+132.5	+3,848	-125.5	-2,887
4	23	25	575	+112.5	+2,588	-131.5	-3,825
5	23	25	575	-112.5	-2,588	-131.5	-3,825
6	23	25	575	-132.5	-3,848	-125.5	-2,887
7	23	25	575	-132.5	-3,848	-125.5	+2,887
8	23	25	575	-112.5	-2,588	-131.5	+3,825
TOTALS							

To calculate the height of 3" chain in a circular chain locker, the calculation can be done with the stowage factor of 3" chain. The stowage factor for 3" chain is 10.3 ft^3 per short ton.

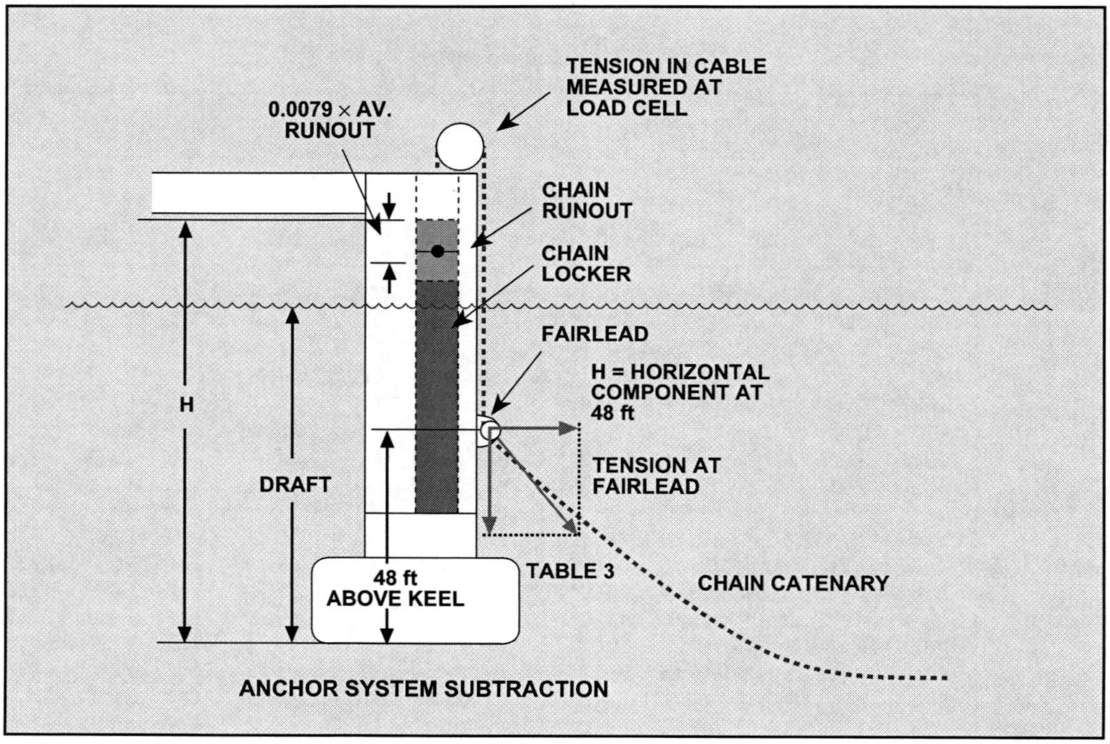

Figure 4.6 Anchor system correction for Rig Type 2A configuration

The equation for a circular chain locker is:

$$\pi \cdot r^2 \cdot h = \frac{\text{Amount of chain in ft} \cdot \text{Weight/ft in air} \cdot \text{Stowage factor}}{2{,}000}$$

The height h can be calculated.

For example: Calculate the height of 700 ft of 3" chain to be added in a 6 ft diameter chain locker.

$$h = \frac{700 \cdot 90 \cdot 10.2}{2{,}000 \cdot 3.14 \cdot 9} \rightarrow h = 11.4 \text{ ft}$$

EXERCISE
Chapter 4

1. Given a chain with a breaking strength of 800,000 lb, determine its:
   a. maximum mooring tension for the survival condition.
   b. proof load.
   c. maximum operating tension (test tension).

2. Use Figures 2.4, 2.11, and 2.13 from Chapter 2 to answer this question for a draft of 70 ft.
   a. What is the total wind force in st for a 30 kt wind from abeam?
   b. What is the abeam current in kt for a current force of 100,000 lb from abeam?
   c. What is the wave drift force in st for a 30 ft significant wave height from abeam?

3. Assume a 30,000 lb anchor has a holding (efficiency) power of 12 in a particular location. How much ground chain with a buoyant weight of 77.36 lb/ft is required if the total bottom part of the anchor system should hold 400,000 lb in soft mud bottom?

4. Calculate the horizontal distance from the fairleader to the touchdown point using 3" chain with a buoyant weight of 77.36 lb/ft. The mooring tension at the fairlead is 300,000 lb. The distance from fairleader to seabed is 650 ft.

5. *This question is optional.* How much chain should be deployed if we want to have 500 ft of ground chain with a mooring line tension of 300,000 lb at the fairleader.

   Water depth = 600 ft
   Buoyed chain wt = 77.36
   Draft = 80 ft
   Fairleader = 48 ft above keel
   Chain wt in air = 89 lb/ft

6. Calculate the total buoyant weight of 2,000 ft of 3" chain including a 30,000 lb anchor. The buoyancy factor for the chain and the anchor is 0.869.

7. Explain the optimum tension.

8. *This question is optional.* Calculate the correction for the vertical moments resulting from deployment of 2,000 ft of chain from a 5 ft diameter chain locker stocked with 4,000 ft of 3" chain.

   Orginal height of top of chain = 85 ft above keel
   Mooring tension = 140 kips
   Chain wt air = 90 lb/ft
   Water depth = 450 ft
   Height fairleader = 48 ft
   Draft = 83.5 ft
   Chain wt buoyed = 78.24 lb/ft
   Stowage factor = 10.2 ft^3/st

CHAPTER 5
Stability at Large Angles—Stability Curves

5.1 INTRODUCTION

Until now, we described the theory of initial statical stability in the vertical, transverse, and longitudinal planes. Initial stability is the stability for small angles up to about 8° to 12°. For small angles, the intersection of the vertical through *B* with the centerline of the rig may be considered to be a fixed point. This point is called the metacenter (*M*).

As a quick reminder, the interaction between *G* and *M* determines the amount of stability.

1. The position of *G* depends on the weight distribution.
2. The position of *M* depends on the rig configuration under the waterline.

For a specific displacement, only the position of *G* can be controlled (within limits) on the rig.

The moment of the righting couple and corresponding arm *GZ* is an indication of the amount of stability. Any vertical change of the position of *G* will increase or decrease the amount of stability because of the change in the length of *GZ*. (fig. 5.1).

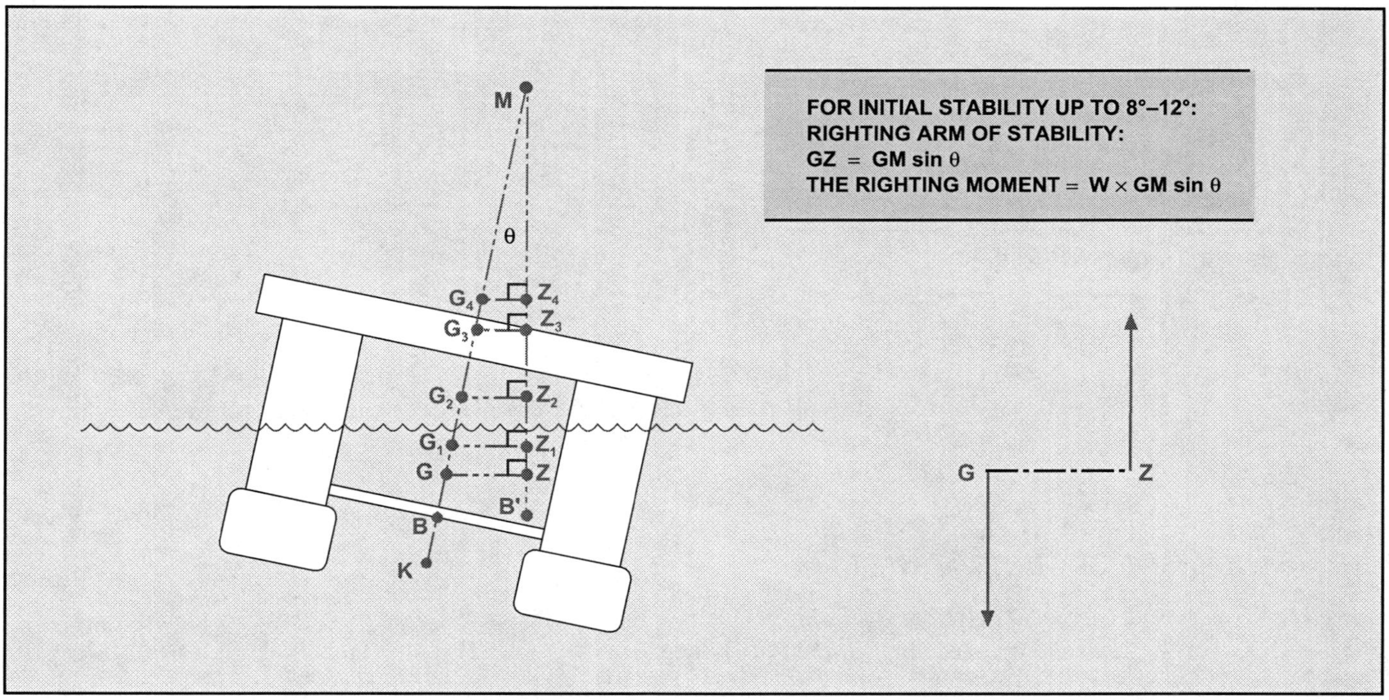

Figure 5.1 Stability couple—Change in the value of the stability arm

5.2 INTACT STATICAL STABILITY FOR LARGE ANGLES

In normal operations, the heel or trim of the rig is considered to be as close as possible to zero. For safety purposes the IMO MODU code specifies that the rig should have enough reserve stability. In other words, the rig should remain safely afloat if, for whatever reason, the rig unexpectedly comes upon a condition with excessive heel and/or trim.

Stability (if positive) is the tendency of a vessel, if heeled or trimmed by an external force, to return to the initial position after the external force is removed.

To comply with the IMO and Classification Society regulations, the rig is theoretically heeled over a range of angles. However, the corresponding righting arms and moments are calculated. For larger angles, *M* can no longer be considered a fixed point. Instead of *M*, the point *N* is used as the intersection of the vertical through *B* and the centerline of the rig. *M* moves away from the centerline.

For calculations for larger angles, the arm of stability is $GZ = GN \sin \theta$. As shown in Figure 5.2, the position of *M* is no longer on the centerline.

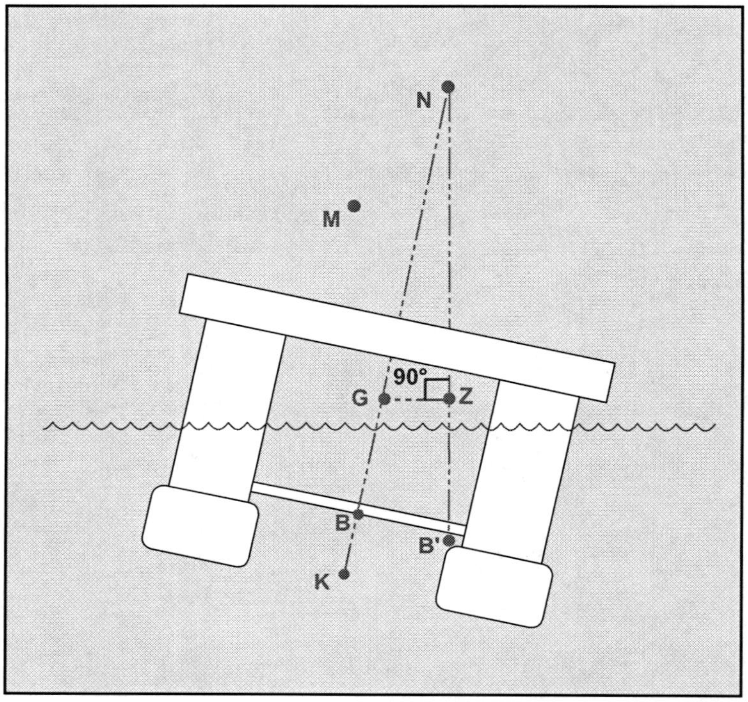

Figure 5.2 Large angle position of M and N

The value of the arm of stability *GZ* is graphically displayed against the angle of heel or trim. The result is the curve of statical stability valid for only

1. the rig as specified.
2. for one displacement.
3. for one value of *G*.

The above is obvious because the formula used is based on the righting moment for:

$$\text{Righting Moment} = W \cdot GZ \text{ and } GZ = GN \sin \theta.$$

$$\text{Righting Moment} = W \cdot GN \sin \theta$$

5.3 THE CURVE OF STATICAL STABILITY

An example of a curve of statical stability for a drilling rig is Figure 5.3. The statical stability curves provide information for one situation on how the stability of the rig will change over a wide range of heel or trim. Stability curves display some very important information such as the

1. maximum arm *GZ* of stability.
2. angle at the maximum arm of stability.
3. initial slope of the curve and the initial *GM* value for stability.
4. range of stability.
5. angle of vanishing stability.

The example of Figure 5.3 shows that the stability of the rig vanishes at ±53°. The arm of the stability increases until the main deck submerges. From this point onward, the waterplane decreases. The maximum arm of stability in the example is ±5.2 ft at an angle of ±34°.

The initial part of the curve is a straight line. It is possible to find the initial *GM* value by drawing the tangent line from the 0° point to the curve. The height of intersection with the vertical line through 57.3° and the tangent line indicate the initial *GM* value. In this case the initial *GM* = ±8.3 ft.

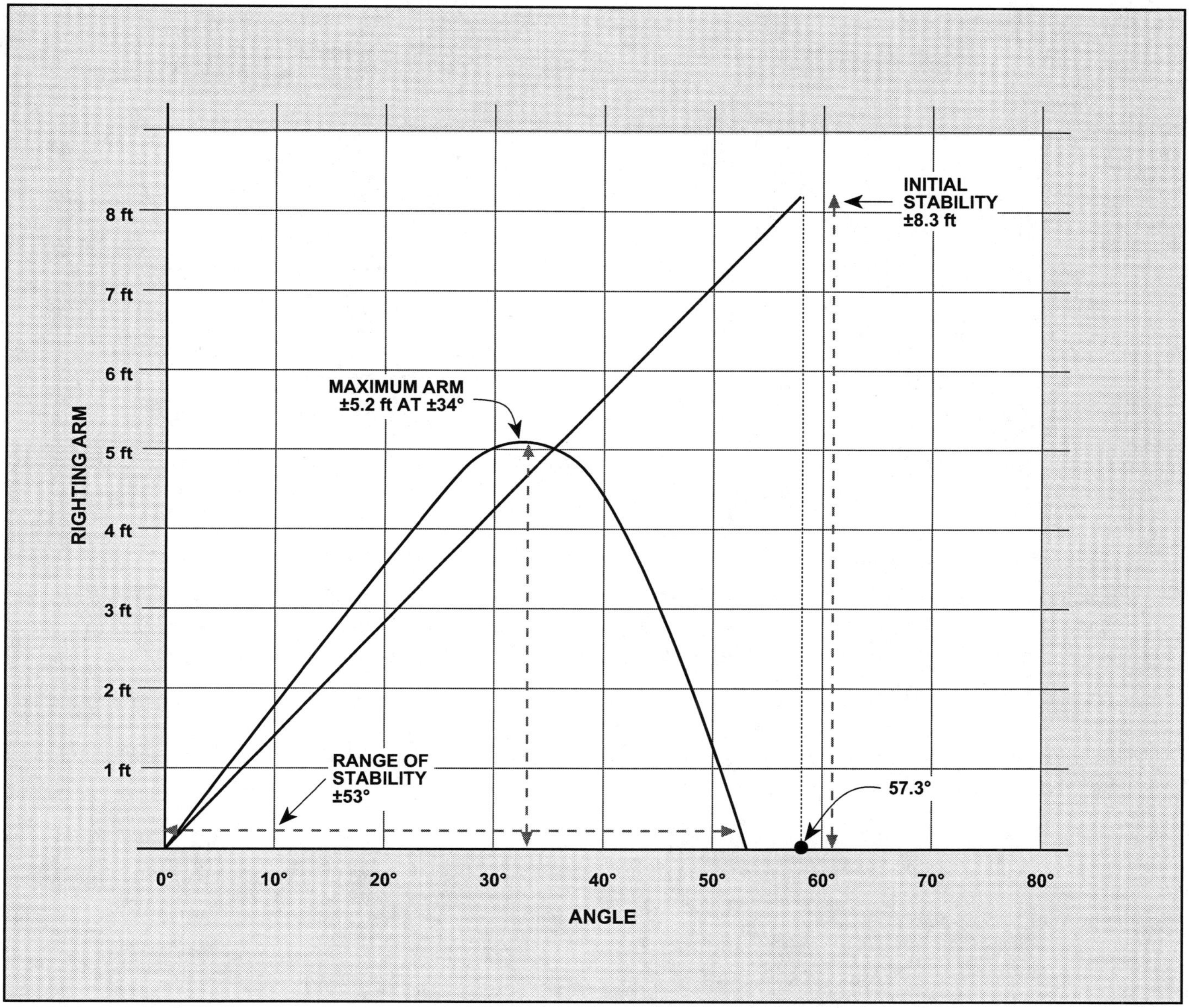

Figure 5.3 Curve of statical stability

5.4 THE EFFECT OF A CHANGE IN WIDTH, CHANGE OF FREEBOARD, AND CHANGE OF G (Fig. 5.4)

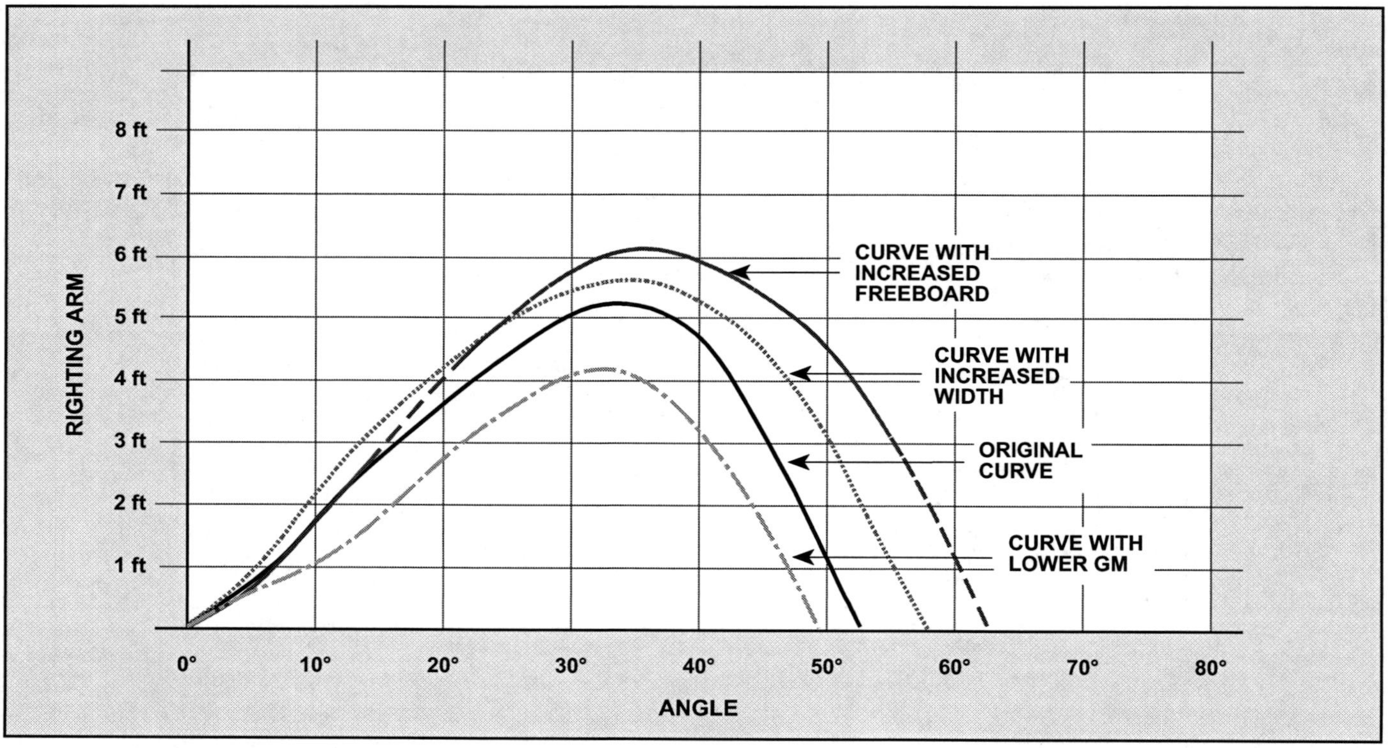

Figure 5.4 **Various curves of statical stability**

1. *The effect of an increase in width.*

   If the width of the vessel from the previous example is increased without changing the draft, freeboard and *KG* value of the stability curve will change. As shown in Chapter 1, the value of *BM* depends largely on the width of a vessel. The result of an increase in width will change the stability as displayed by the ·········· curve because of

   a. an increase in *BM*, *GM*, and the arm of stability *GZ*.
   b. an increase in the range of stability.
   c. a decrease in the down-flooding angle.
   d. the value of *KB* remains the same.

2. *The effect of an increase in freeboard.*

   If for the same vessel the freeboard is increased without changing the *KG* value and the width of the vessel, the stability will change as displayed by the — — curve because of

   a. an increase in the down-flooding angle.
   b. an increase in the range of stability.
   c. a decrease in *KB* value because of a lighter draft.
   d. an increase of *GM* and *GZ* because larger down-flooding angle causes an increase in the waterplane.

3. *The effect of a decrease in GM.*

   If for the same vessel only the value of *GM* decreases, the stability will change as displayed by the —·— curve because of

   a. a decrease of the arm of stability *GZ* over the entire range.

5.5 THE STATICAL STABILITY CURVE FOR THE ANGLE OF LOLL

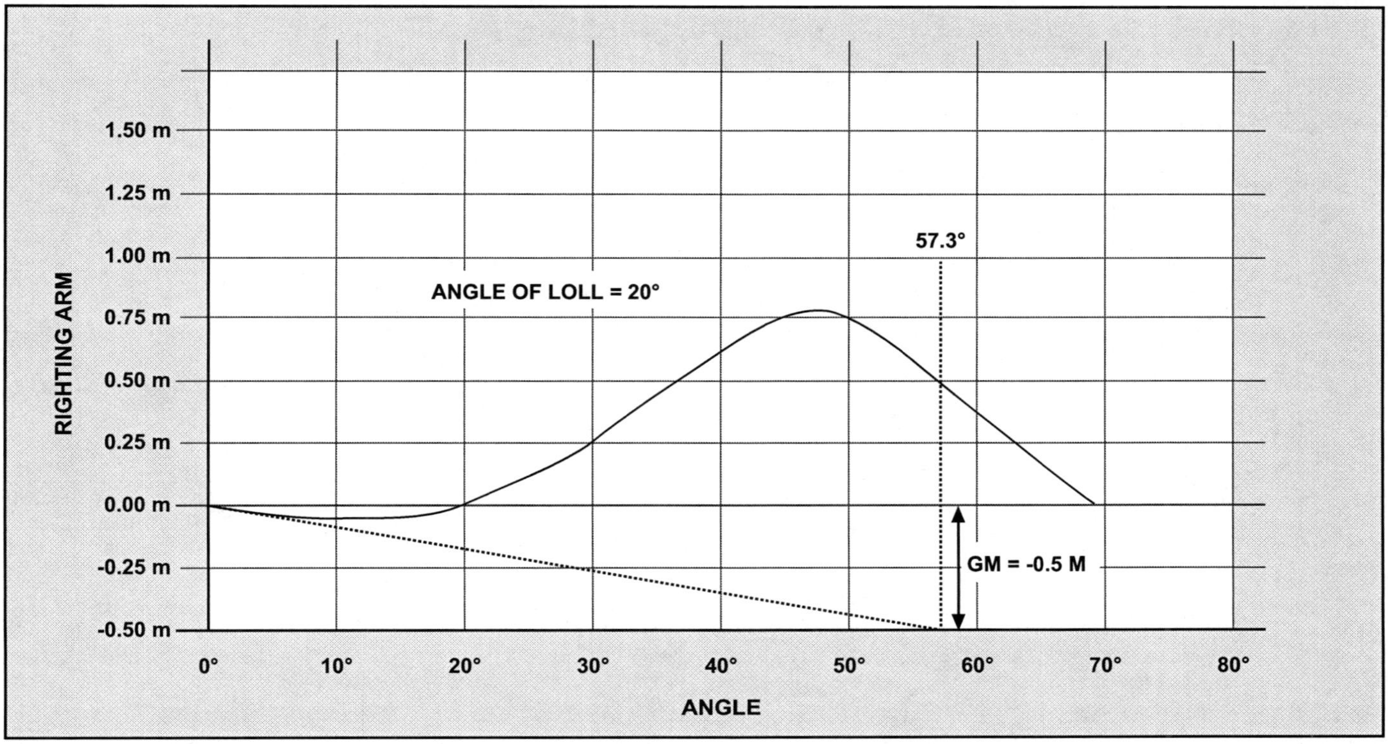

Figure 5.5 **Stability curve for initial negative *GM* and angle of loll**

As explained in Chapter 1, in theory a rig with a negative initial *GM* will capsize. In actuality, this will not happen. The negative couple inclines the rig until the center of buoyancy moves into a position vertically under *G*. At this angle, the rig will have zero stability. This angle is called the angle of loll. The angle of loll can be on either side. This means that the rig can flip over to the other side at any moment. This is a very dangerous situation. An angle beyond the angle of loll will increase the waterplane. Any increase of angle caused by an outside force will produce a positive stability couple because *B* moves further to the low side and *M* moves up. The direction of the stability couple is now positive.

The example of Figure 5.5 shows that the

a. angle of loll is 20°.
b. range of positive stability is from 20° to 70°.
c. initial *GM* = -0.5 m.
d. max righting arm is 0.77 m at about 47°.

5.6 EXAMPLES OF STABILITY CURVES FOR VARIOUS TYPES OF VESSELS

The shapes of stability curves for vessels and MODUs are quite different. This is shown in Figure 5.6. The stability curves as shown are typical for a general cargo vessel, a passenger vessel, and a semisubmersible drilling rig.

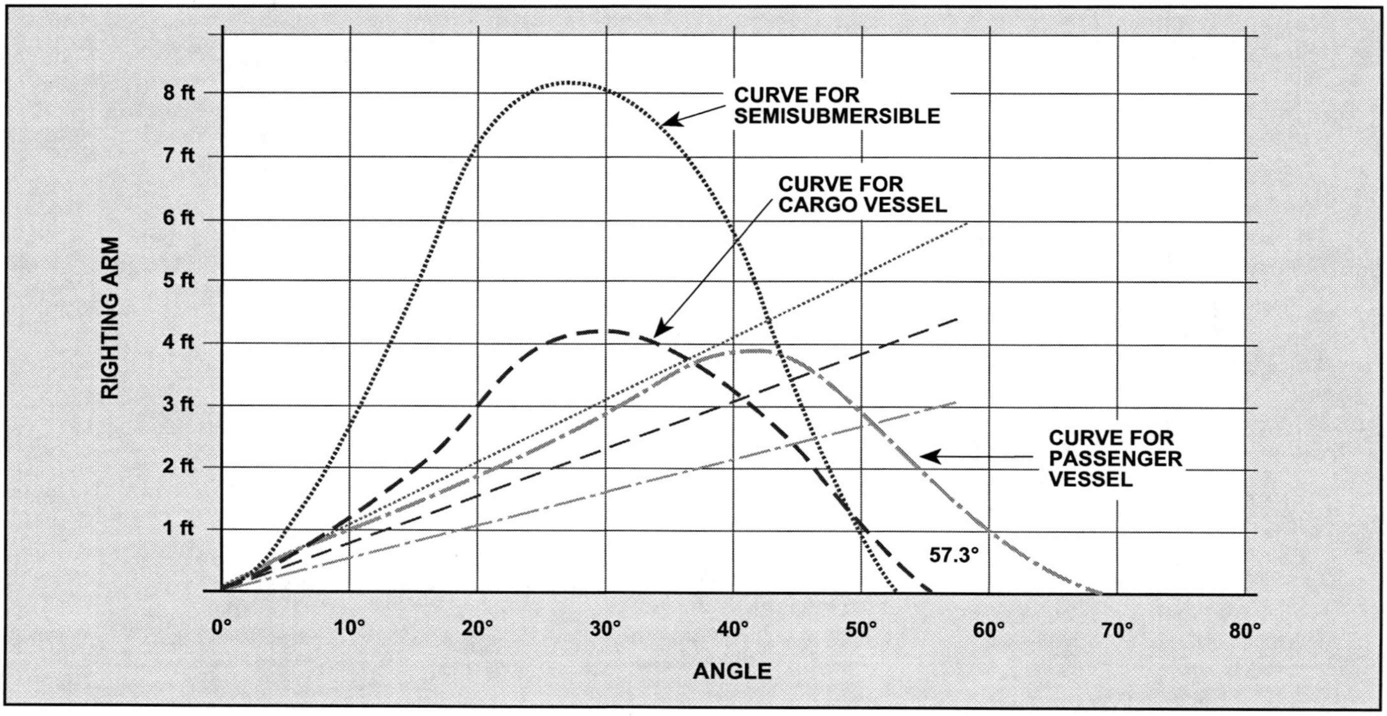

Figure 5.6 Static stability curves for three types of vessels

1. *The maximum arm of the stability couple* is reached at the top of the curve. For vessel-shaped drilling units, this is the moment where the main deck starts to immerse. From thereon the waterplane area starts to decrease and the buoyancy decreases rapidly. However, on semisubmersibles, the waterplane area still increases on the high side. Sometimes the deckhouses may still provide for additional buoyancy.

   What are the values for the maximum arm of stability in Figure 5.6?
   a. The passenger vessel ________ ft
   b. The cargo vessel ________ ft
   c. The semisubmersible ________ ft

2. *The angle of inclination at maximum arm of stability* is reached at the top of the curve. An increase of freeboard will increase the angle of downflooding. The effect of the increase of freeboard is explained in section 5.4.

   What are the values for the angle at maximum arm of stability in Figure 5.6?
   a. The passenger vessel __________ °
   b. The cargo vessel __________ °
   c. The semisubmersible __________ °

3. *The initial amount of stability* expressed in the value of *GM* for small angles depends on the initial slope of the stability curve. To determine the initial value of *GM*, draw a vertical line at 57.3°. Draw a second line as the tangent to the curve from the 0° point. The height of the intersection of the two lines on the stability arm scale gives initial *GM* value. *GM* initial = 3 ft. for the passenger vessel. For the first 8° to 12° the stability curve is plotted as a straight line from base point *B* to about 8°.

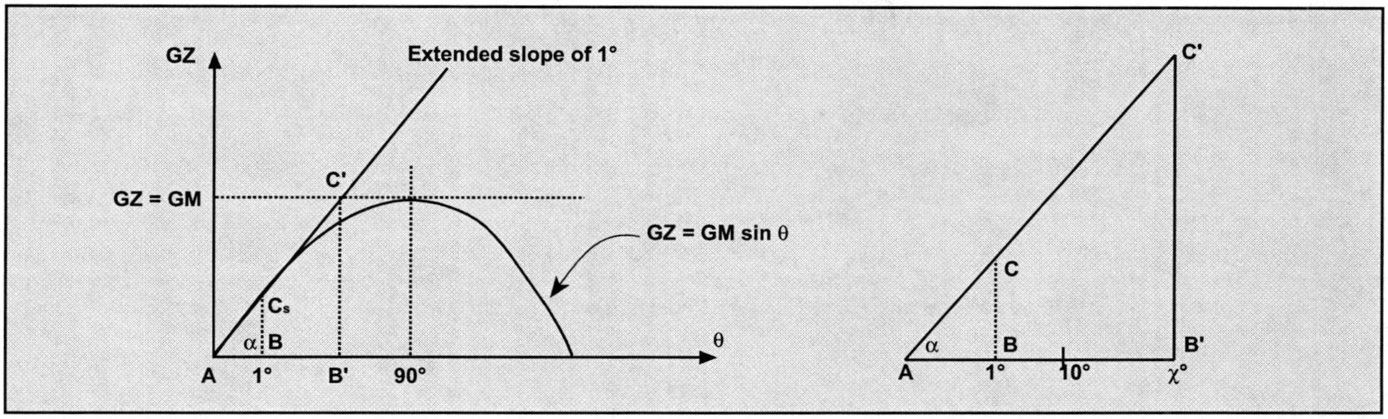

Figure 5.7 Derivation of GM versus 57.3°

The explanation of the *GM* value versus 57.3° is displayed below with the formula and in Figure 5.7.

Assume GM is constant at all angles → $GZ = GM \sin \theta$

$$\tan \theta \text{ (or slope)} = \frac{CB}{AB} = \frac{C'B'}{AB'}$$

$$AB' \text{ (angle)} = \frac{AB \cdot C'B'}{CB} \quad \text{Name } AB' \text{ as } \chi°$$

$$\chi° = 1° \cdot \frac{C'B'}{CB} \text{ or}$$

$$\chi° = 1° \cdot \frac{GM}{GM \sin 1°}$$

$$\chi° = \frac{1°}{\sin 1°}$$

$$\chi° = 57.3°$$

4. *The range of stability* relates to the angle where no positive stability is left. The range of stability has only academic value because the curve represents a static situation for each angle. In the real world, the situation will be different:

- Cargo will start to move at large angles.
- The down-flooding angle will be reached before the curve zero crossing.
- Dynamic forces such as wind and wave are not included.
- Once the rig reaches the down-flooding angle, the rig will start to sink. Water will enter into the chain locker and/or other nonwatertight compartments.

5.7 THE CROSS CURVES OF STABILITY

The statical curve of stability is valid for only one value for displacement and one *KG* value. This is not very useful. To prevent long calculation for each new combination of displacement and *KG* values, the *cross curves of stability* provide the possibility to construct stability curves for practically any combination of *KG* and displacement values.

The first calculations are done with one righting arm for one angle and one *KG* value plotted against the displacement scale from lightship weight to maximum load condition. Subsequent calculations are done with the same *KG* value for a range of angles with intervals of 5° or 10°.

Cross curves of stability with a value of KG = 0 is preferrable because the correction is always negative. This leaves no room for errors.

Because cross curves of stability are only valid for one *KG* value, a correction is required for any other *KG* value.

If *G* is the position of the center of gravity for the cross curves and *G′* the actual position of the center of gravity, the correction to apply is:

For Figure 5.8 → G'Z' = GZ + G'A
(a lower KG produces a larger arm of the stability couple)

and because GA = GG' sin θ

G'Z' = GZ + GG' sin θ

For Figure 5.9 → G'Z' = GZ + GA
(a higher KG produces a smaller arm of the stability couple)

and because GA = GG' sin θ

G'Z' = GZ − GG' sin θ

If actual KG is lower than the KG used for the curves, the correction is positive (+)

If actual KG is higher than the KG used for the curves, the correction is negative (−)

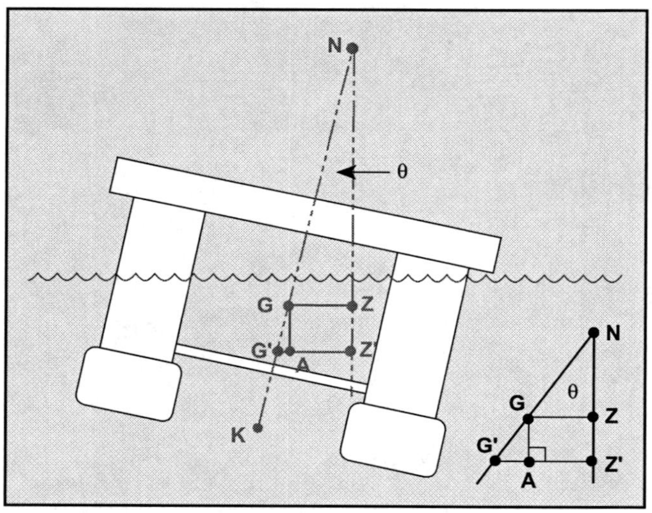

Figure 5.8 GG' correction for stability cross curves

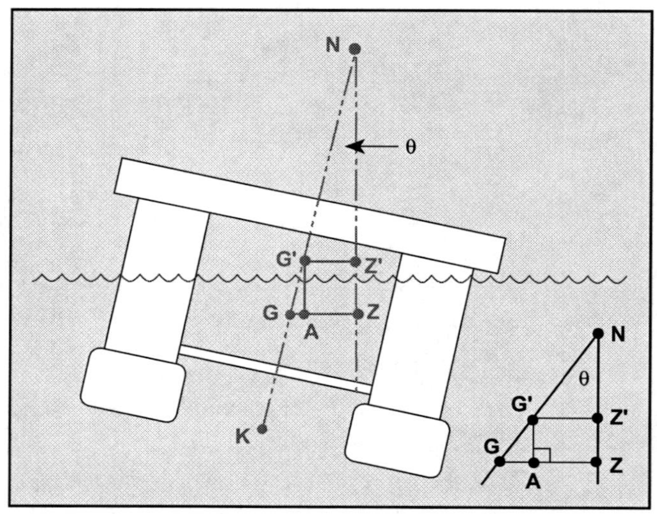

Figure 5.9 GG' correction for stability cross curves

Figure 5.10 is an example for the metric system on a Rig Type 2 with an assumed *KG* value of 15 m. Other rigs may use cross curves of stability with an assumed *KG* = 0. The example of Figure 5.11 displays the Imperial system with an assumed *KG* value of 0 ft.

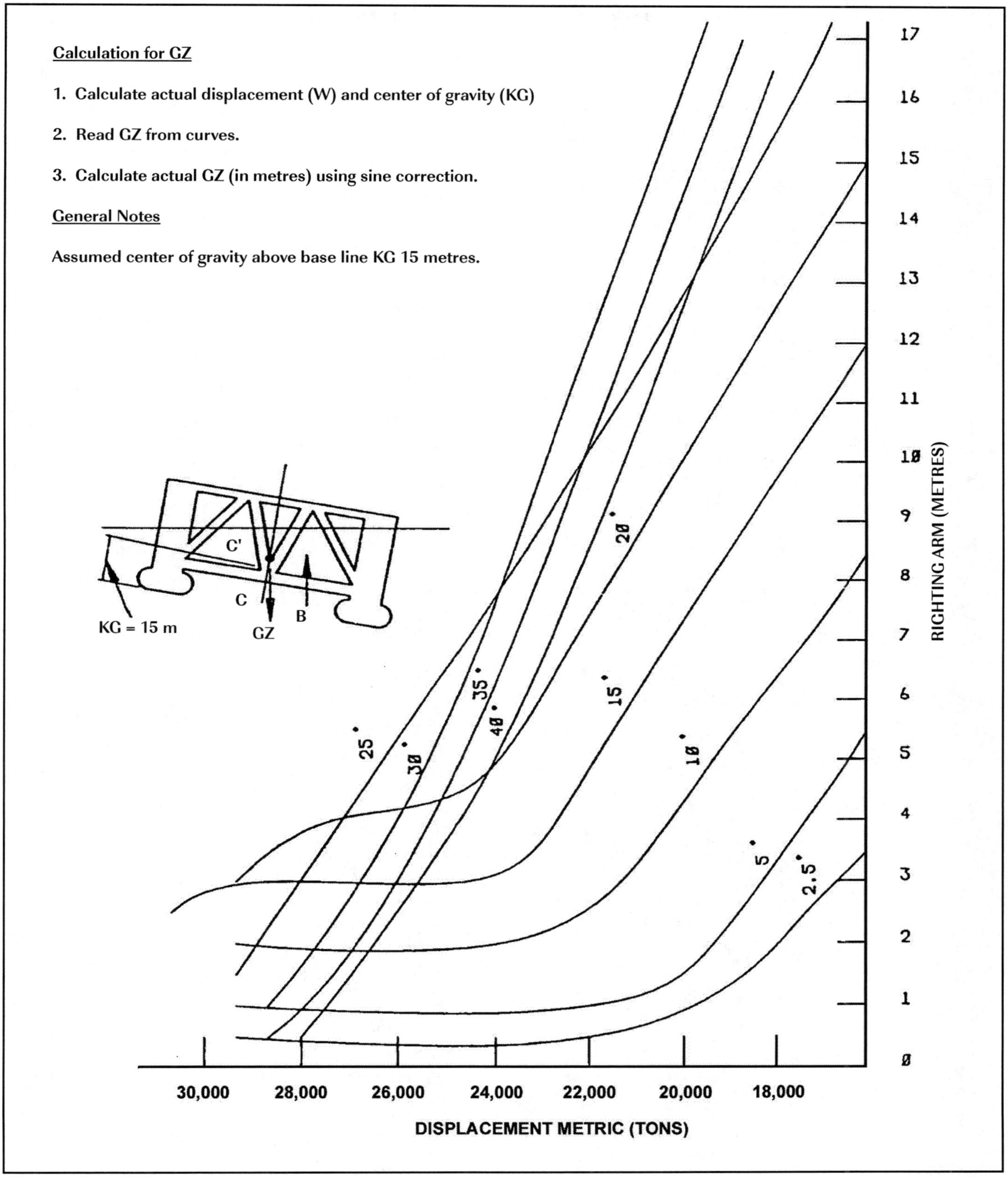

Figure 5.10 Cross curves of stability metric system—Rig Type 2 with KG value = 15 m

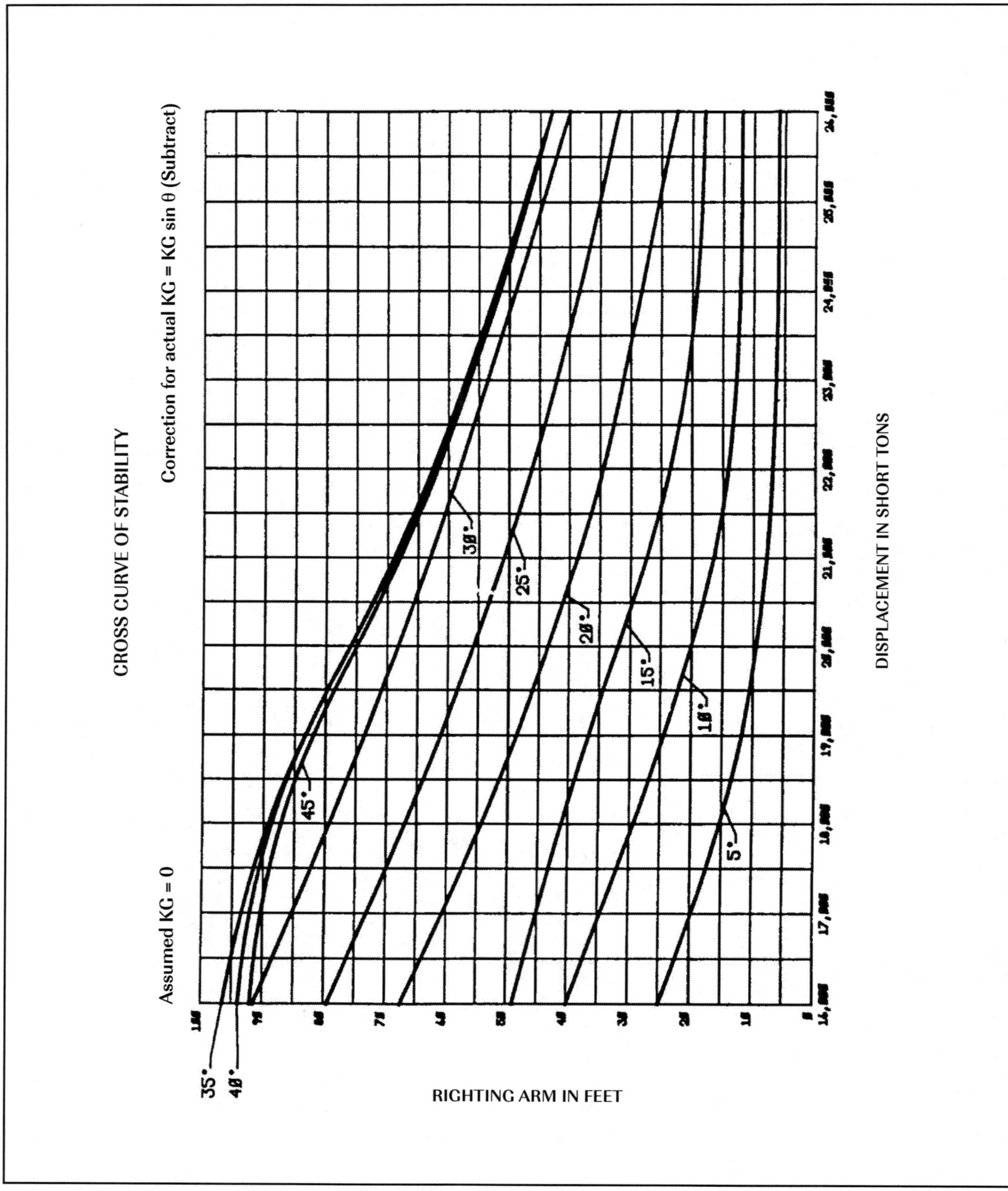

Figure 5.11 Cross curves of stability—Imperial system KG value = 0 ft

5.8 SUMMARY OF USE OF CROSS CURVE OF STABILITY

To summarize how to use the cross curves of stability, follow steps 1 through 5:

1. Calculate the displacement and *KG* value.
2. Insert a vertical line through the actual displacement.
3. Take the perpendicular value to find the righting arm for each angle.
4. Apply the correction $GG' \sin \theta$ for the actual *KG* value for each angle.
5. Draw the stability curve.

Example. Find the actual static stability curve.

The stability curve of a vessel with a $KG = 3$ ft is known as displayed in the example. The actual $KG = 3.34$ ft. Plot the actual static stability curve. See Figure 5.12.

sin 0° = 0	sin 15° = 0.25	sin 30° = 0.50
sin 40° = 0.64	sin 60° = 0.37	

GG = 3.34 ft − 3.00 ft = 0.34 ft

The correction is negative because KG increased.

Angle	GG' × sin θ	Correction
0°	0.34 × 0.00	−0.00 ft
15°	0.34 × 0.25	−0.09 ft
30°	0.34 × 0.50	−0.17 ft
40°	0.34 × 0.64	−0.22 ft
60°	0.34 × 0.87	−0.30 ft

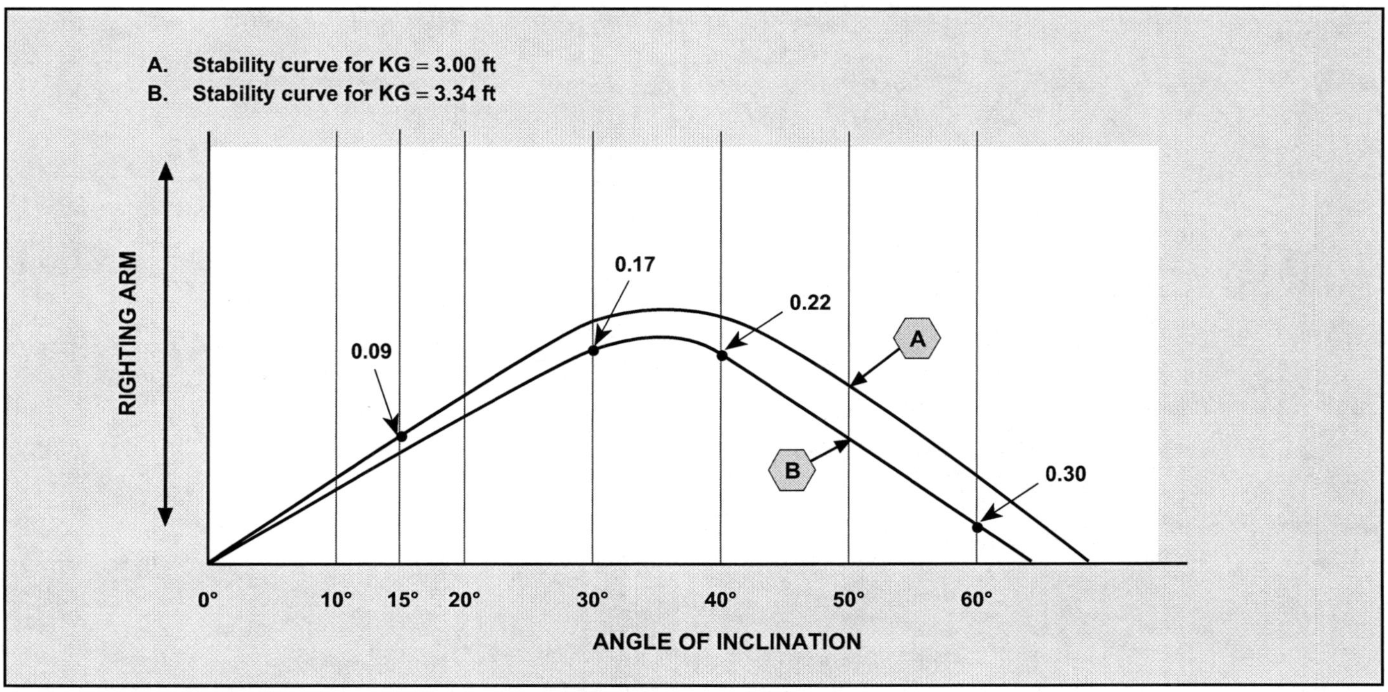

Figure 5.12 Static stability curve for actual KG value of 3.34 ft.

5.9 THE CURVE OF RIGHTING MOMENTS

The statical stability curve from Figure 5.3 displays the arm of stability *GZ* against the inclination. We can go one step further and construct the righting moment curve. The value of *GZ* for each specific angle is now multiplied by the weight of the displacement (*W*) and plotted against the angle of heel or trim. The result is a curve that looks exactly the same because W is a constant factor in the formula $W \times GZ$ (fig. 5.13, left side).

5.10 DYNAMIC STABILITY

The static stability curves with the arm of stability or the righting moment only displays a statical righting arm or moment for a specific angle. The statical curve does not provide any information on the amount of energy required arriving at a specific angle.

Dynamic stability is more than just a moment. It is the work done in inclining a vessel or rig. The work done is the force times the distance. If the energy required to incline the rig in st•ft or t•m to each angle is plotted against the angle, the curve represents the energy curve. It shows that it is quite easy to incline the rig to a very small angle, but very difficult to incline the rig to a large angle.

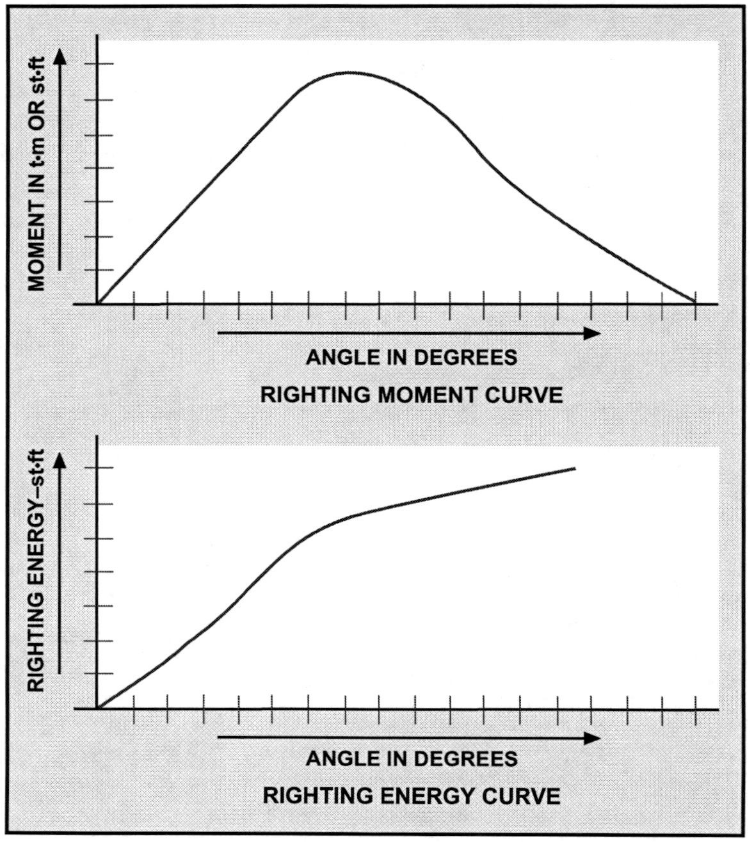

Figure 5.13 Righting moment curve versus righting energy curve

One step further is to take the righting moment curve from Figure 5.13 and take each individual smallest angle and add the corresponding moments (of each angle), and, up until the angle of inclination, we find the total energy. The sum of all the moments represents the area under the curve. The work required to heel or trim the rig is proportional to the area under the curve of the statical stability. Therefore, we can say that dynamic stability at any angle is equal to the area under the curve of righting moments, which is displacement multiplied by the arm of stability (fig. 5.14).

Dynamic stability = W · area under the stability curve.

5.11 THE STABILITY CURVE AND THE WIND HEEL CURVE

The dynamic stability of a rig is important for the Classification Societies and Flag State authorities to verify that the stability criteria are in accordance with the IMO MODU Code. The IMO regulations impose a specified wind heeling moment on the righting moment curve. The result should confirm that the rig has enough reserve stability to remain in a safe condition without getting in a down-flooding angle at a too early stage. The wind heel moment is calculated the same way as other moments by application of a force against the area.

The formula used to calculate the overturning wind heel moment is:

Overturning moment (t·m or ft·st) = $P \times A \times H$

P = The wind pressure in t/m^2 or st/ft^2
A = The area in m^2 or ft^2
H = Height of the center of application above the waterline

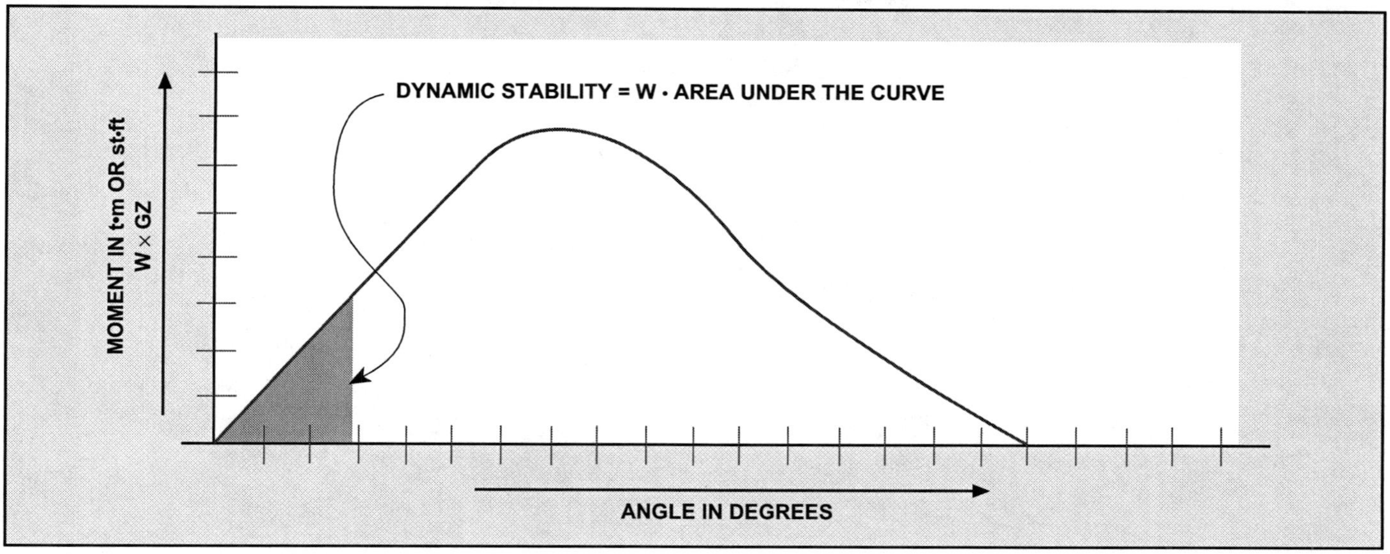

Figure 5.14 Dynamic stability—Area under the curve

In addition to the wind force on the rig structure, columns, and deckhouses, the underside of the deck of a semisubmersible will be exposed to the wind force when the rig starts to incline. The wind force and corresponding overturning moment will not be constant. After increasing because of the additional underdeck effect, the wind force curve will decrease because the height factor decreases.

The energy curve of wind in st•ft or t•m is superimposed on the dynamic stability curve. The rig heels over until the amount of the wind energy equals the energy absorbed by the rig. In the example of Figure 5.15, this angle is about 9°. This point is called the first intercept.

The remaining reserve energy for stability is the area above the wind heel curve. The vanishing point with the wind heel curve is reduced to about 63°. The example shows one case how the authorities evaluate the stability behavior of the rig by imposing various storm conditions.

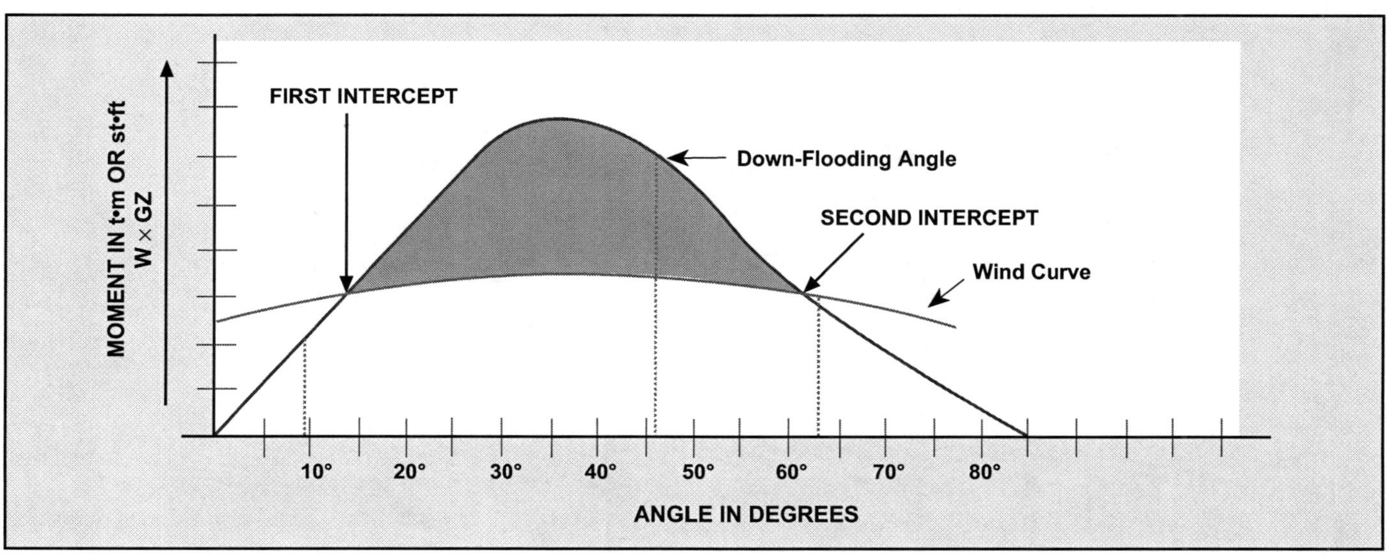

Figure 5.15 Stability curve and wind curve

5.12 WIND CRITERIA

In most cases, the wind criteria to evaluate and analyze the stability for MODUs for intact and damaged stability are:

1. 70 knots (36.0 m/s) for operational and transit conditions.
2. 100 knots (51.5 m/s) for survival conditions.
3. 50 knots for the damaged stability condition.

5.13 INTERNATIONAL STABILITY REGULATIONS

As early as 1960, the first rules and regulations concerning stability for MODUs were developed by ABS. The IMO developed the Code for Construction and Equipment of MODUs which includes a section on the inclining test, righting moments and heeling moments curves, and intact and damaged stability criteria.

Each Classification Society followed up with their own rules based on the IMO Code. It goes beyond this course to go into all the details but, basically, all classification societies use similar criteria with minor differences.

5.14 INTACT AND DAMAGE STABILITY CRITERIA

The criteria used by the classification societies and flag states are based on the IMO Code for Construction and Equipment of MODUs for two situations:

1. *Intact stability*
   - ❖ The rig operating condition.
   - ❖ No uncontrolled ingress of water.
   - ❖ The ballast system is fully operational.
   - ❖ The rig has enough stability to sustain severe weather conditions within the safety margins with the wind criteria:
     a. 70 knots for operational conditions.
     b. 100 knots for survival conditions

2. *Damaged stability*
   - ❖ The rig sustained damage.
   - ❖ The remaining stability and buoyancy should be enough to keep the rig afloat.
   - ❖ Wind criteria: 50-knots wind.
   - ❖ Inclination MODU Code <25° (but ABS uses <17°).
   - ❖ Damage based on:
     One compartment for jackup.
     One compartment for a semisubmersible, but if a watertight boundary lies within the damage zone, damage to two compartments is involved.
   - ❖ Extent of the damage:
     a. For a jackup between watertight bulkheads horizontal penetration 1.5 m.
        Vertical from baseline upward without limit.
     b. For a semisubmersible on columns horizontal extend 1.5 m.
     c. Vertical extent of 3 m between any level from 5 m above to 3 m below the waterline. Including damage to a horizontal watertight bulkhead if the boundary of the horizontal flat lies within the damaged zone (2 compartment damage) (fig. 5.16).

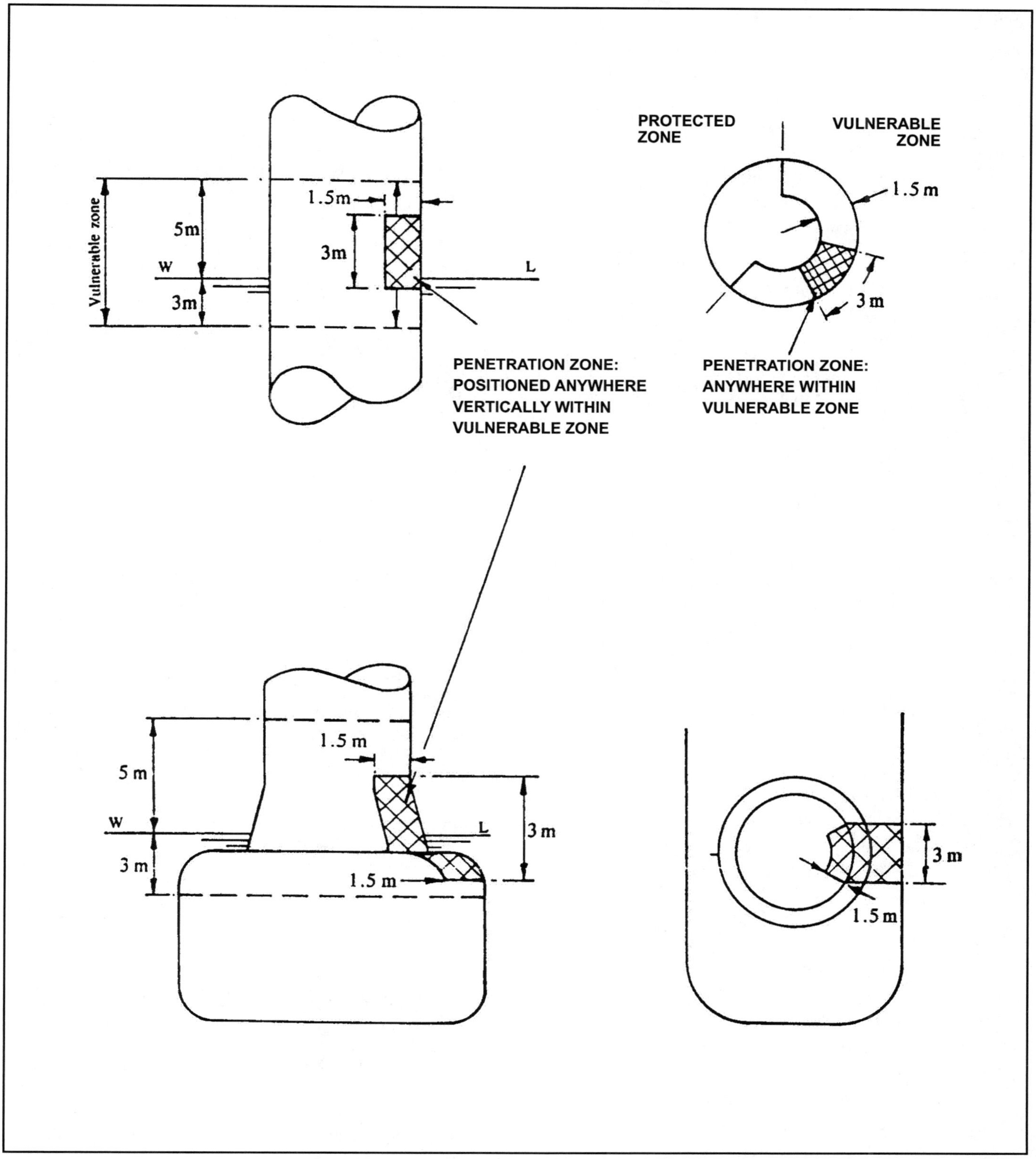

Figure 5.16 Example with penetration criteria for a semisubmersible

5.15 DAMAGE STABILITY

Damage stability is the uncontrolled or unintentional flooding of a compartment. The cause of damage stability can be:

- ❖ collision.
- ❖ grounding with damage to one or more tanks.
- ❖ free flooding through an open manhole.
- ❖ burst waterline.
- ❖ failure of the ballast system.
- ❖ fire extinguishing water ingress.
- ❖ structural failure.

A damaged stability will change any one or a combination of the rig conditions:

1. Heel and/or trim.
2. The mean draft.
3. The stability.
4. The reserve buoyancy.
5. The anchor chain tensions.

5.16 THE DOWN-FLOODING ANGLE

The analysis of the stability curve includes the down-flooding angle. The down-flooding angle is the angle when the nonwatertight openings on the main deck start to submerge. The result is further uncontrolled flooding. Theoretically, the rig cannot be saved if it reaches the down-flooding angle. The down-flooding angle for most rigs takes place before the vanishing point of the stability curve. In Figure 5.15, the down-flooding angle is at about 46°.

5.17 THE RIGHTING MOMENTS

The righting moments and stability in each mode of operation should meet the following criteria:

1. The righting moment should be positive over the entire range of angles from upright to the second intercept.
2. The area under the stability curve from first intercept to the down-flooding angle or second intercept, whichever occurs first, is used to analyze the amount of required stability. On semisubmersibles, the down-flooding angle always occurs before the second intercept.
3. The area under the righting moment curve to the second intercept or down-flooding angle, whichever occurs first, should not be less than 40% in excess of the area under the wind heeling moment curve for self-elevating units or 30% for column-stabilized units. This is referred to as the energy ratio rule. Expressed in formula using Figure 5.17.

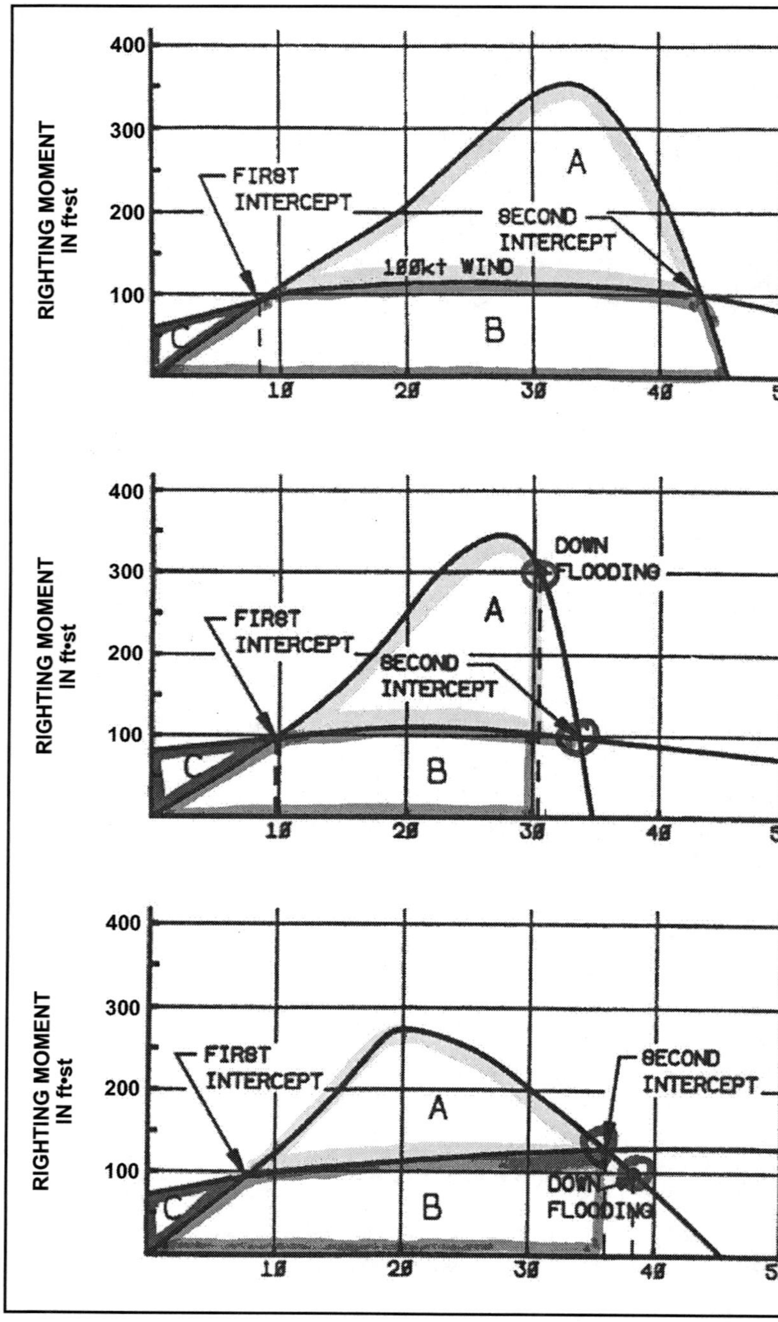

Figure 5.17 Stability righting moment curves and areas under the curve

For jackups: Area A + Area B > 1.4 (Area B + Area C)

For semisubmersibles: Area A + Area B > 1.3 (Area B + Area C)

4. In general, the operations draft and survival draft for semisubmersibles are not the same. A smaller draft results in a larger down-flooding angle and a larger area under the stability curve. When eventually the tubulars penetrate the water level, the waterplane area increases. The result is an increase of the righting moment. The stability analysis is therefore calculated for various drafts.
5. In the damaged condition, the regulations require that the rig remains afloat with positive stability including a 50-kt wind from any direction. This means that the righting moment curve, after the sustained damage, must be above the 50-kt wind curve. For the given draft, the inclination under the damaged condition must be less than the down-flooding angle or second intercept.

As mentioned before, each classification society or flag state may use more stringent rules than the IMO Code. The details are specified in the MOM.

5.18 MAXIMUM ALLOWABLE VCG OR *KG* CURVES

To maintain the minimum required stability criteria responsibly, the naval architect conducts stability calculations with the maximum allowable *KG* values for various drafts and conditions. This ensures that the rig will comply with the design criteria for operation and survival conditions and always has enough reserve stability and buoyancy.

After completion of all the calculations, the maximum allowable VCG values are plotted against the draft and/or displacement. The maximum allowable VCG values are part of the information in the MOM. The OIM and barge supervisor are responsible to maintain the VCG values at all times within the limits as displayed on the maximum allowable VCG curves or tables. Figures 5.18, 5.19, and 5.20 are typical examples of such curves and tables.

5.19 THE EFFECT OF ICING ON STABILITY

When working in subzero temperatures, the effect of icing on the stability of a rig can be considerable. The effect is the same as if a load with a high *KG* value is added to the rig. The center of gravity (G) will rise and the stability will decrease. If the total weight of the ice is known, the righting arm is reduced with GG' sin θ. If the displacement is W, the total righting moment is reduced with:

$$GG' \sin\theta \cdot W$$

ALLOWABLE VCG VALUES

Draft (m)	Draft (ft)	Survival Conditions m (ft)	Drilling Conditions m (ft)	Moving Conditions m (ft)	Intermediate Temporary Conditions (Passing Through) m (ft)
8.50	27.89			44.63 (146.42)	44.63 (146.42)
9.00	29.53			40.99 (134.48)	40.99 (134.48)
9.50	31.17			37.35 (122.54)	37.35 (122.54)
9.70	31.82			35.89 (117.76)	35.89 (117.76)
10.00	32.81				33.71 (110.60)
10.50	34.45				31.53 (103.43)
11.00	36.09				29.34 (96.26)
11.50	37.73				27.16 (89.09)
12.00	39.37				24.97 (81.92)
12.50	41.01				24.75 (81.19)
13.00	42.65				24.53 (80.46)
13.50	44.29				24.30 (79.73)
14.00	45.93				24.08 (79.00)
14.50	47.57				23.86 (78.27)
15.00	49.21				23.64 (77.54)
15.50	50.85				23.41 (76.81)
16.00	52.49	23.19 (76.08)			23.19 (76.08)
16.50	54.13				22.94 (75.26)
17.00	55.77				22.69 (74.44)
17.50	57.41				22.44 (73.61)
18.00	59.06				22.19 (72.79)
18.50	60.70				21.94 (71.97)
18.75	61.52		21.81 (71.56)	21.81 (71.56)	21.81 (71.56)
19.00	62.34		21.81 (71.54)	21.81 (71.54)	21.81 (71.54)
19.50	63.98		21.80 (71.52)	21.80 (71.52)	21.80 (71.52)
20.00	65.62		21.79 (71.49)	21.79 (71.49)	21.79 (71.49)
MAXIMUM ALLOWABLE DRAFT IS 20.00 m (65.62ft)					

NOTES:-

1. **All units are in metres and (feet)**
2. **AVCG values exclude free surfaces correction which must be added to the calculated VCG.**
3. **The VCG values given are controlling over the deck load weight in Table 1.2.**
4. **Deballasting to a Survival Draft should begin when the wind or rig motion exceeds the amounts described in Section 3.4.2.2 or when waves threaten to strike the deck at drafts of 18.75 m (61.52 or more as described in the same section. This defines the change from "moderate seas" to "storm seas."**

Figure 5.18 Example of the VCG values of the Rig Type 1

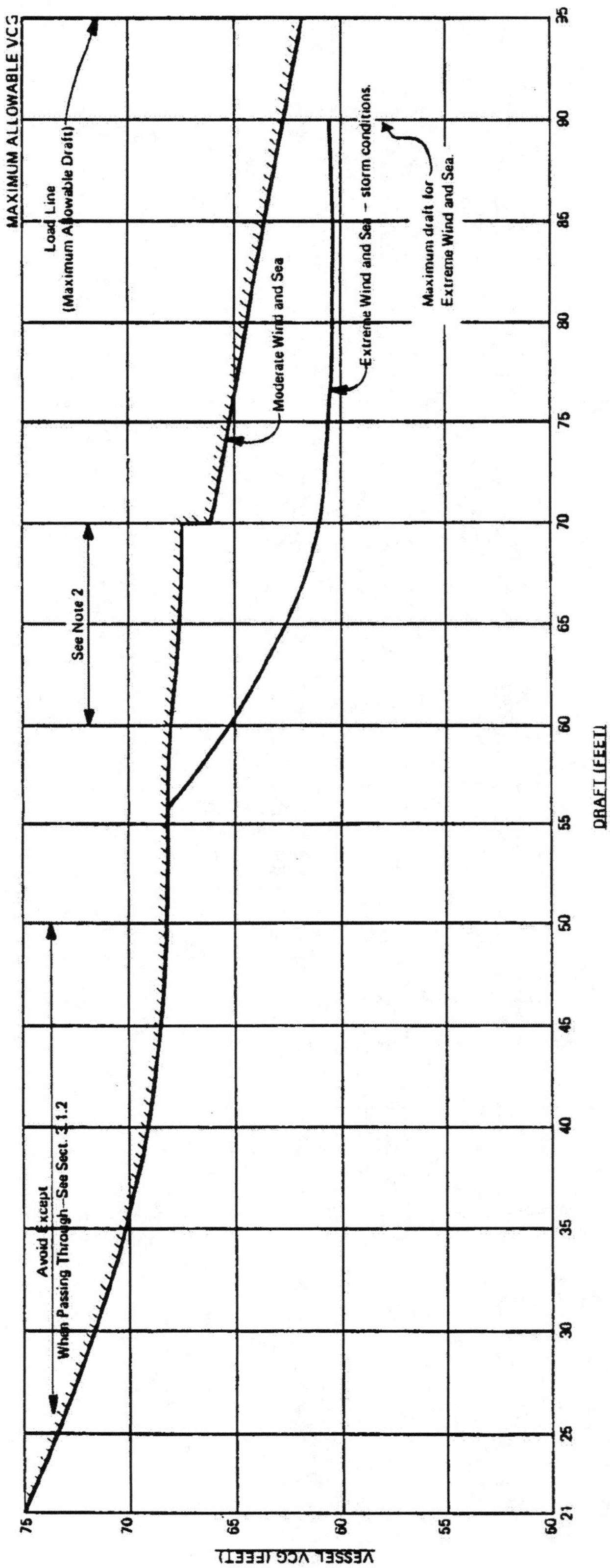

NOTES:
1. Actual unit VCGs must always be less than the allowable VCG outlined by the hatched line.
2. Drafts between 60 and 70 feet are not permitted except when passing through or in extreme weather. (See Section 3.1.2)
3. When wind exceeds 70 knots, VCG must be less than solid line. (Extreme wind and sea – storm conditions)
4. Allowable VCGs have maximum operating free surface effect as described in Section 3.2.4.5

Figure 5.19 Example of VCG curve for Rig Type 2

Table 2-7 – Maximum Allowable KG vs. Displacement (Continued)

Displacement (metric tons)	Maximum Allowable KG (m)		
	Intact[1]	Damage	
		Transit/Drilling[2]	EWT[3]
98,000	14.90	15.35	14.98
100,000	14.83	15.31	14.98
101,710	14.77	15.27	14.98

Notes:

1. Intact severe storm condition, wind speed 51.5 m/s (100 knots).
2. Damage conditions applicable to operating and transit conditions.
3. Damage conditions for extended well test mode; n/a means not applicable at the given displacement.

Figure 2-1 – Maximum Allowable KG Curve

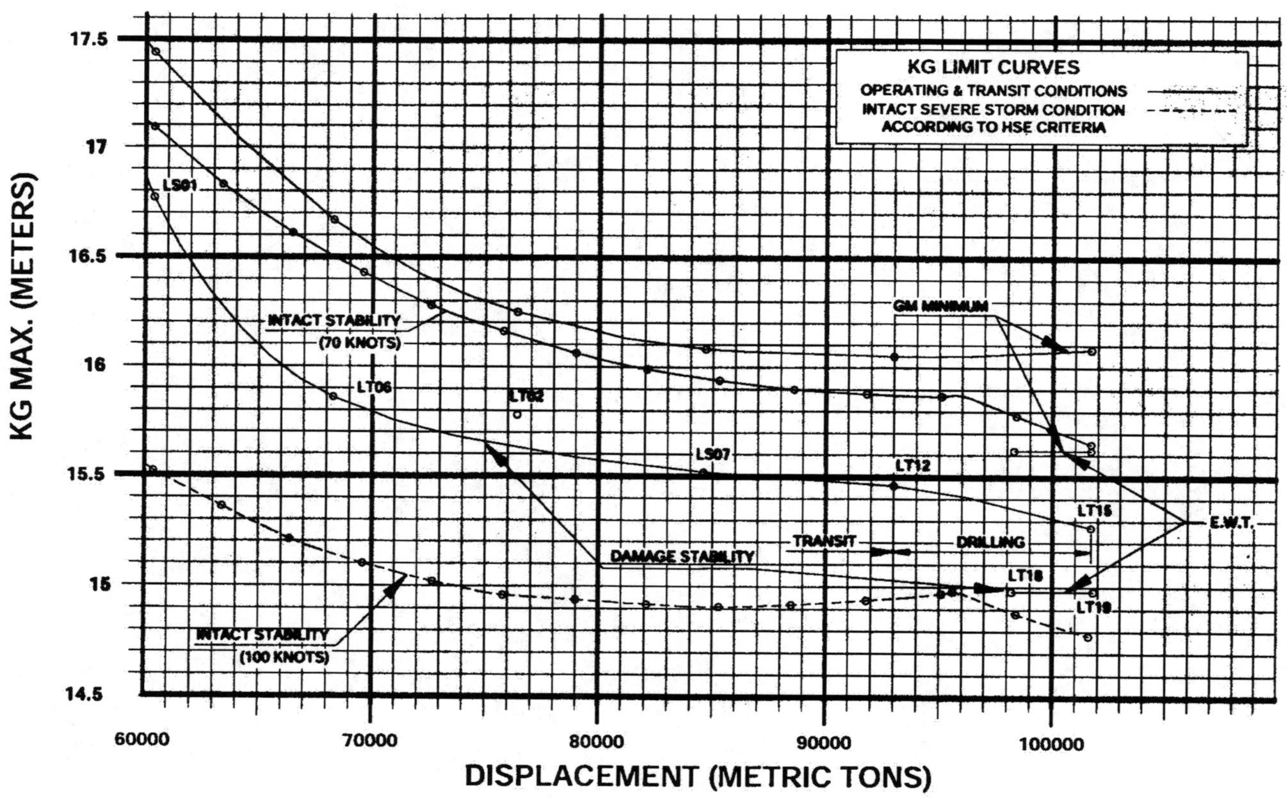

Figure 5.20 Example of the KG curves for Rig Type 4

To find the actual stability curve, the correction is deducted from the original curve. Figure 5.21 shows an example of the effect of icing on the stability curve.

As shown in Figure 5.22, in some marine operations manuals like on Rig Type 2A, tables with ice or snow thickness versus weight and an ice load factor, allows a quick calculation to determine the effect of the ice load on the stability of the rig.

5.20 THE EFFECT OF A WEIGHT SHIFT ON THE STABILITY CURVE

If a weight (W) is shifted over a distance (d) from port to star-board, the heeling moment can be calculated with:

$$W \cdot d \cdot \cos\theta$$

If the calculation is done over a range of angles, the heeling righting moment curve can be superimposed on the stability righting moment curve. The result will show that the angle will be caused by the weight shift (fig. 5.21).

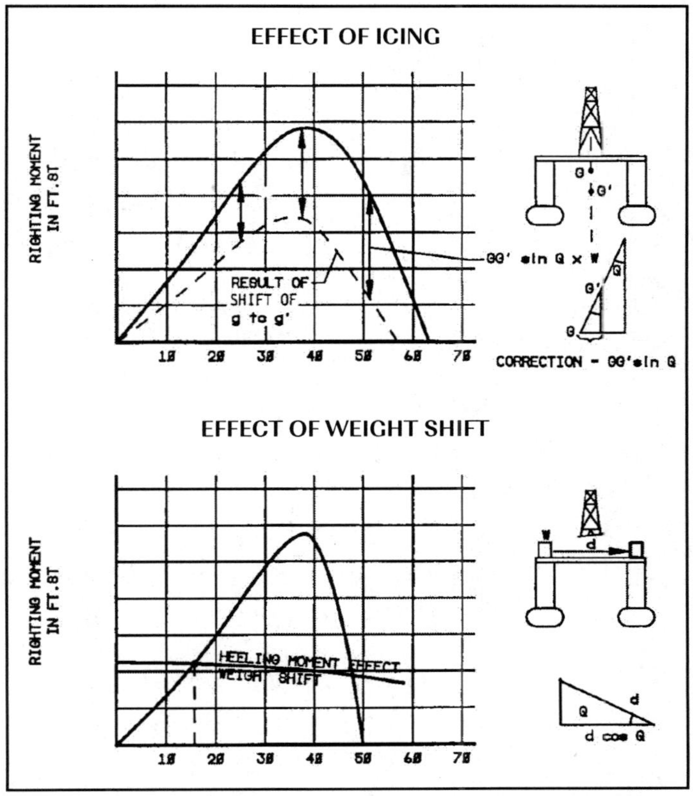

Figure 5.21 Effect of icing and weight shift on the righting moment curve

TABLE 11.2
RIG TYPE 2A
ICE AND SNOW LOAD MAIN DECK AND ABOVE

Ice Snow Thickness (in.)	Ice Load (ST)	Snow Load (ST)
1	115	15
2	230	30
3	345	45
4	460	60
5	575	75
6	–	90
7	–	105
8	–	120
9	–	135
10	–	150

Operating Draft VCG = 144 ft ABL
Survival Draft VCG = 148 ft ABL

TABLE 11.1
RIG TYPE 2A
SEA SPRAY ICE LOAD BELOW MAIN DECK

Ice Thickness (in.)	Ice Load[a] Factor (ST/ft)
1	0.81
2	1.62
3	2.43
4	3.24
5	4.05

Operating Draft VCG = 108 ft ABL
Survival Draft VCG = 104 ft ABL

[a]This ice load is determined by multiplying ice load factor times exposed vertical height of rig below main deck and above wave wash zone. See Form 4, which assumes a wave-wash zone of 10 ft.

RIG TYPE 2A
ICE AND SNOW LOAD CALCULATION

Item	Thickness Ice/Snow (in.)	Ice Factor	Difference (120 – Draft) (ft)	Ice/Snow[a] Load (ST)	VCG (ft)	Vertical Moment (ft-ST)
Table 11.1 Below Main Deck						
Table 11.2 Main Deck and Above		X	X			
Draft ______			Totals[b]			

[a]Ice load below main deck is determined by multiplying ice factor times difference (120 – Draft).
[b]Enter totals into loading report, Form 1, Page 8, Miscellaneous Loads.

Figure 5.22 Tables to calculate the effect of icing on stability

EXERCISE
Chapter 5

1. With the cross curves from chapter 5, Figure 5.11, draw the curve of righting moments for a displacement = 20,605 st and *KG* = 65 ft. Given the heeling moment from a 100-kt wind is 37,100 st•ft at zero degree angle, and 80,780 st•ft at 28.7° heel. What is the down-flooding angle? Superimpose the wind heeling moments and estimate the angle of steady heel caused by the wind.

2. Draw a curve of righting moments for a displacement = 25,938 st and *KG* = 55 ft. Given the heeling moment caused by a 100-kt wind is 26,175 st•ft at zero degree and 53,215 st•ft at 19.4° heel. What is the angle of down-flooding? Superimpose a curve of wind heeling moments and estimate the angle of steady heel caused by the wind.

3. Superimpose on the righting moment curve from question #1 a heeling moment curve caused by the flooding of an empty ballast tank PB3.

   PB# is centered 110 to port of the rig's TCF.
   Use the tank specification from Chapter 3, Table 3.6.
   Draw the curve as if *KG* remains the same.

4. Draw a single righting moment curve (without the wind curve from question #1) after the flooding of tank PB3 that represents the rig's stability after the flooding. Subtract the cosine correction curve representing the effect of the flooding from the original curve.

5. In the previous questions #3 and #4, no correction was made for the change of *KG* caused by the flooding. If the correction of *KG* is applied to the calculation, would the angle of list be greater or less than the results calculated curves from questions #3 and #4. Clarify your answer.

6. An ice load is evenly distributed across the rig, so its only effect on the rig's stability will be a change in *KG*. TRUE or FALSE? Clarify your answer.

7. Why is the rig prohibited from operating or staying at range of drafts as shown in the shaded areas in Figure 5.18?

8. The survival draft for a semisubmersible is less than the operations draft. What effect does this have on the stability curve?

9. What is the transverse down-flooding angle for a semisubmersible with a width of 77 m and an air gap of 16 m?

CHAPTER 6
Damage Stability

6.1 DEFINITION OF DAMAGE STABILITY

A vessel or rig is in a damaged stability condition whenever uncontrolled flooding of any compartment develops or exists. In the marine world, this is also called bilging based on a holed compartment below the waterline.

6.2 CAUSES OF DAMAGE STABILITY

A damage stability condition may be caused by:

1. Collision
2. Grounding
3. Free flooding between one or more compartments through a normally closed opening such as a watertight door or manhole.
4. Flooding through a burst waterline, leaking valve, or leaking flange.
5. Cracks in structural members followed by flooding.
6. Failure of the ballast system combined with uncontrolled filling of ballast tanks.
7. Icing if the buildup of ice weight results in a neutral or negative stability.
8. Water buildup in a compartment from water used to extinguish a fire.

6.3 CONSEQUENCES OF DAMAGE STABILITY

The effect of damage stability will include one or any combination of the following changes:

1. Change in heel.
2. Change in trim.
3. Change in mean draft.
4. Change in reserve buoyancy.
5. Change in anchor tensions.
6. Elimination of part of the ballast system caused by flooding in the pump room.

It is important to realize that the analysis of a damaged stability condition is based on the theory that the vessel or rig remains afloat with positive stability after the damage to enable successive, controlled counteraction to save the unit from sinking or capsizing.

6.4 PERMEABILITY

Permeability in marine terms is the percentage of water a compartment can contain after it has been flooded. A void space without any equipment could be filled with a volume of water equal to the volume of the void space. Its permeability is close to 100%. The permeability of a pump room with a considerable amount of built-in equipment may be only 60%. A higher value of permeability results in a higher amount of loss of buoyancy. To calculate the effect of flooding, the permeability of a compartment must be known.

6.5 RESERVE BUOYANCY AND DAMAGE STABILITY

Reserve buoyancy is the total volume of watertight space above the waterline. In the intact condition, the reserve buoyancy can be described as the amount of weight that could be added to a rig in upright condition before all watertight spaces are fully submerged. For a semisubmersible, in most cases, the main deck is the uppermost watertight deck. In a damaged stability condition, a rig without reserve buoyancy will immediately go into the progressive flooding condition and sink.

6.6 WATERTIGHT INTEGRITY

If all the reserve buoyancy on a MODU was confined to one space without partitions, damage to the waterline would result in total flooding. The MODU will lose all reserve buoyancy at once and sink.

To prevent such a major loss of reserve buoyancy, the structure of the MODU is divided into multiple watertight compartments below and above the waterline. This division into watertight compartments is called the watertight integrity. Because it is not feasible to build an unsinkable vessel or MODU, the number of watertight compartments are intended to allow one or two compartments to endure damage. With one or two compartments flooded, the rig should remain afloat with enough reserve buoyancy and stability.

In Chapter 8, the various regulatory requirements for MODUs for damaged condition were discussed. The OIM on board should be aware of the damage stability criteria and limitations as prescribed by the Classification Society and Flag State authorities. He should be familiar with the operations and survival limitations of the unit. A very important part is that the watertight integrity of the unit is maintained at all times. Openings in the watertight compartments may be one or more of the following appliances:

1. Watertight doors
2. Balls in the goosenecks of ventilation pipes
3. Watertight hatches
4. Manholes
5. Inlet and discharge valves
6. Ventilation intakes and outlets
7. Chain locker openings

Watertight doors, hatches, and manholes should be kept closed except for passage or specific maintenance and inspection tasks. Other appliances such as gooseneck balls, fail-safe valves, and remote-controlled ventilation closures should be functional at all times.

In case of damage to a compartment with a watertight door or hatch in open position, flooding of an adjacent compartment will, in most cases, result in an inclination beyond the down-flooding angle and the rig will likely capsize or sink.

6.7 CALCULATION OF DAMAGE STABILITY

The calculation to determine the effect on the stability and inclination of a vessel or MODU after sustaining damage can be done two ways:

1. With the added weight method.
2. With the lost buoyancy method.

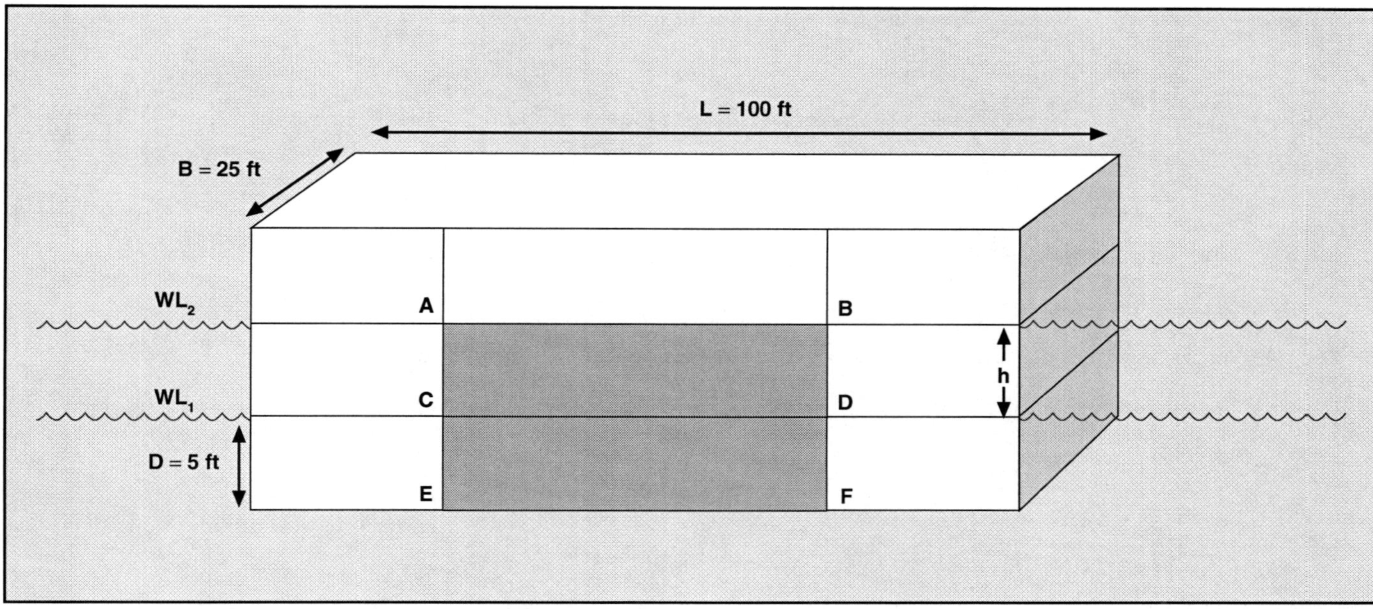

Figure 6.1 **Box-shaped barge added weight method**

To explain the two principles to calculate the stability after a one compartment damage, a simple box-shaped barge is used with the following dimensions. See Figure 6.1.

- Length =100 ft. Breadth = 25 ft. Depth = 15 ft.
- The barge is divided into three compartments.
- The side tanks are 25 ft long and the center tank is 50 ft long.
- Initial draft = 5 ft.
- The initial KG = 8 ft.
- The center tank is damaged and floods.
- Calculate the damaged draft and new GM.

6.8 THE ADDED WEIGHT METHOD

The added weight method is mostly used to determine the effect of damaged stability because it is easy to understand and to calculate.

With the added weight method, the calculation is based on the assumption that the flooded compartment is intentionally filled and the weight of the seawater is added as an additional load.

The calculation to find the new draft and GM for the barge as shown in Figure 6.1 after the center tank is damaged and fills with seawater is:

Normally the hydrostatic properties on board provide information to calculate the sinkage with TPI.

For the box-shaped example, the calculation for sinkage is done by conversion of the added water volume into weight.

1. Find sinkage

After the compartment CDEF is flooded, the barge settles to a new waterline WL_2.

The volume of the increased displacement equals the volume in the flooded compartment or:

Volume between WL_1 and WL_2 = the volume of section ABFE.

$100\text{ ft} \cdot 25\text{ ft} \cdot h = (50\text{ ft} \cdot 25\text{ ft} \cdot 5\text{ ft}) + (50\text{ ft} \cdot 25\text{ ft} \cdot h)$

$h = 5\text{ ft} \rightarrow$ sinkage $= 5\text{ ft} \rightarrow$ the new draft $= 10\text{ ft}$

2. The next step is to find the added weight.

The volume of displacement before flooding at $WL_1 = 100\text{ ft} \cdot 25\text{ ft} \cdot 5\text{ ft} = 12{,}500\text{ ft}^3$

The volume of displacement after flooding at $WL_2 = 100\text{ ft} \cdot 25\text{ ft} \cdot 10\text{ ft} = 25{,}000\text{ ft}^3$

$$\text{Initial displacement} = \frac{12{,}500\text{ ft}^3}{31.25\text{ ft}^3/\text{st}} = 400\text{ st (because } 1\text{st} = 31.25\text{ ft}^3)$$

$$\text{Final displacement} = \frac{25{,}000\text{ ft}^3}{31.25\text{ ft}^3/\text{st}} = 800\text{ st} \rightarrow \text{added weight} = 400\text{ st at a KG} = 5\text{ ft}$$

3. Find new KG

Initial KG barge = 8 ft

$$GG' = \frac{w \cdot d}{W + w} = \frac{400\text{ st} \cdot (8\text{ ft} - 5\text{ ft})}{800\text{ st}} = 1.5\text{ ft}$$

New KG = 8 ft – 1.5 ft = 6.5 ft

4. Find free surface effect

$$\text{Free surface correction} = \frac{I}{V} = \frac{L \cdot B^3}{12V} = \frac{50\text{ ft} \cdot 25\text{ ft}^3}{12 \cdot 25{,}000\text{ ft}^3} = 2.6\text{ ft}$$

5. Find new GM_{wet}

$KM = KB + BM$

$$KB = \frac{\text{Draft}}{2} = \frac{10\text{ ft}}{2} = 5\text{ ft}$$

$$BM_{barge} = \frac{I}{V} = \frac{L \cdot B^3}{12V} = \frac{100\text{ ft} \cdot 25\text{ ft}^3}{12 \cdot 25{,}000\text{ ft}^3} = 5.2\text{ ft}$$

$KM = 5\text{ ft} + 5.2\text{ ft} = 6.2\text{ ft}$

$GM_{dry} = KM - KG = 6.2\text{ ft} - 6.5\text{ ft} = 3.7\text{ ft}$

$GM_{wet} = GM_{dry} - FSE = 3.7\text{ ft} - 2.6\text{ ft}$

$GM_{wet} = 1.1\text{ ft}$

The initial GM calculated with the same formula as above is 4.92 ft.

The reduction in GM from 4.92 ft to 1.1 ft because of the flooded compartment is substantial.

6.9 THE LOST BUOYANCY METHOD

The same barge is used to calculate the result of the flooded compartment with the lost buoyancy method (fig. 6.2).

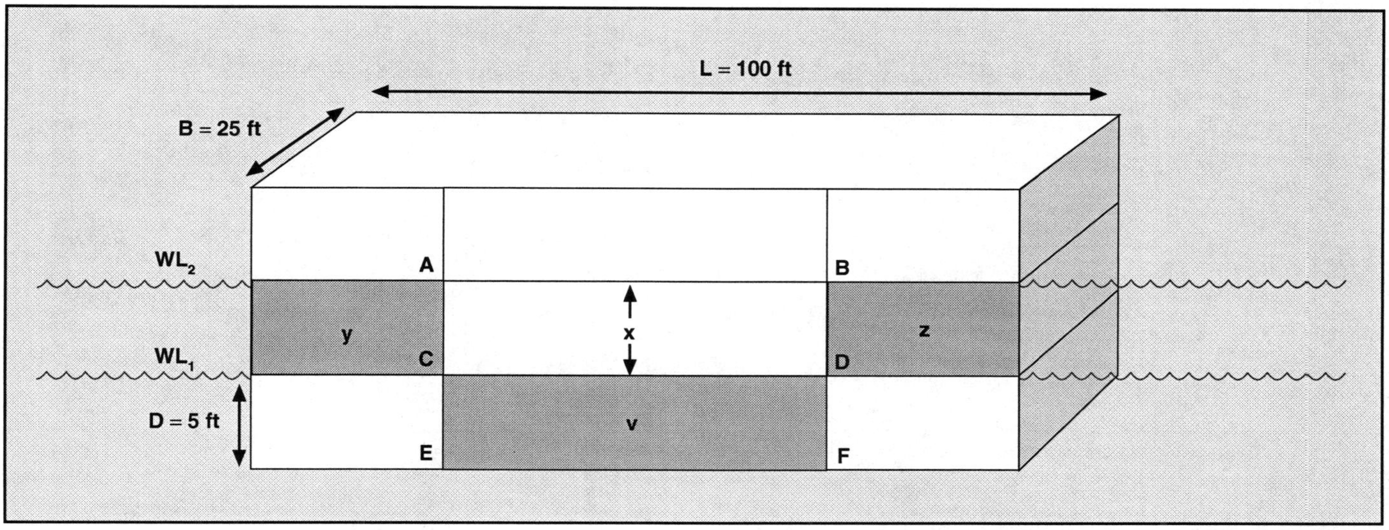

Figure 6.2 The lost buoyancy method

With the lost buoyancy method again, the center tank is holed and free communication exists between the sea and the damaged compartment. The loss of buoyancy is compensated with the increase in draft. In other words, the volume of the lost buoyancy is replaced by an equal amount of increase in underwater volume.

The volume of buoyancy that is lost is CDEF (this does not include ABDC since this was reserve buoyancy)

1. Calculate increase in draft. The volume of buoyancy that is gained is the volume between WL_1 and WL_2 minus the volume ABCD since this section no longer provides buoyancy.

So, volume CDFE = volume between WL_1 and WL_2 minus volume ABDC.

$50 \text{ ft} \cdot 25 \text{ ft} \cdot 5 \text{ ft} = (100 \text{ ft} \cdot 25 \text{ ft} \cdot x \text{ ft}) - (50 \text{ ft} \cdot 25 \text{ ft} \cdot x \text{ ft})$, where $x = 5$ ft.

2. Calculate KB, KM, and BM. Like in the previous calculation KB = 5 ft.

Because the buoyancy lost equals the buoyancy gained, the total displacement did not change. This is true because, with the principle of the lost buoyancy, the flooded compartment is no longer part of the barge.

$$BM = \frac{I}{V} = \frac{50 \text{ ft} \cdot 25 \text{ ft}^3}{12 \cdot 12{,}500 \text{ ft}^3} \text{ or } BM = 5.2 \text{ ft}$$

$$KM = KB + BM = 5 \text{ ft} + 5.2 \text{ ft}$$

$$KM = 6.2 \text{ ft}$$

3. Calculate GM.

$$GM = KM - KG = 6.2 \text{ ft} - 8 \text{ ft}$$

$$GM = 2.2 \text{ ft}$$

The initial KG did not change because, with the lost buoyancy method, the flooding is not considered as an added weight. There is also no FSE.

This time the initial GM is reduced to 2.2 ft which is different than the 1.1 ft from the added weight method.

6.10 COMPARISON BETWEEN ADDED WEIGHT AND LOST BUOYANCY METHODS

The difference in GM value between the added weight and lost buoyancy methods can be explained. With the added weight method, the total weight of the barge increased from 400 st to 800 st. With the lost buoyancy method, the weight of the barge is considered to remain the same at 400 st.

For small angles, the righting moment (W • GM • sin θ) is a measure for initial stability and not the GM value.

For the added weight method, the righting moment is:

RM = 800 st · 1.1 ft · sin θ

For the lost buoyancy method, the righting moment is:

RM = 400 st · 2.2 ft · sin θ

So both righting moments are the same.

It proves that both methods are correct.

Note: the initial RM = 400 st · 2.2 ft · sin θ

6.11 THE STABILITY CURVE FOR DAMAGE STABILITY

To analyze the effects of the damage stability condition at large angles, various flooding conditions are computed. For each condition, the heeling moment curve is superimposed on the righting moment curve as explained in Chapter 8. Figure 6.3 is an example of such a combination of curves.

The heeling moment curve as explained is produced by:

Heeling moment due to flooding = w · d · cos θ

W = the weight of the added water.

In this case, the list will be approximately 8° and the angle where all stability vanishes is approximately 51°.

The down-flooding angle depends on the draft and the watertight integrity of the rig and is most probably somewhere between 35°and 40°.

The area A is the remaining righting moment curve after the flooding. Subtracting the value of the heeling moment curve from the righting moment curve produces the final curve (fig. 6.4).

6.12 DAMAGE CONTROL PROCEDURE

It would be unrealistic to assume a rigid step-by-step procedure could be applied to every damaged stability condition. The procedures to follow are rig specific. Consult the damage control section of the rig's marine operations manual with the details of the damage control procedures which, for example, includes thruster flooding and, if installed, the use of the secondary ballast system.

However, in general, the basic procedures to follow are:

1. Before starting any counteraction, determine the extent and the location of the damage. This may include actions such as visual inspection and soundings of the suspected compartment(s).
2. If flooding occurs because of an external damage by collision, the first action to be taken can be decided upon fairly quickly. Internal flooding by failure of a pump, sea chest seals, or a broken line may require more time.
3. Organize to muster the crew to the designated areas to ensure nobody is missing.

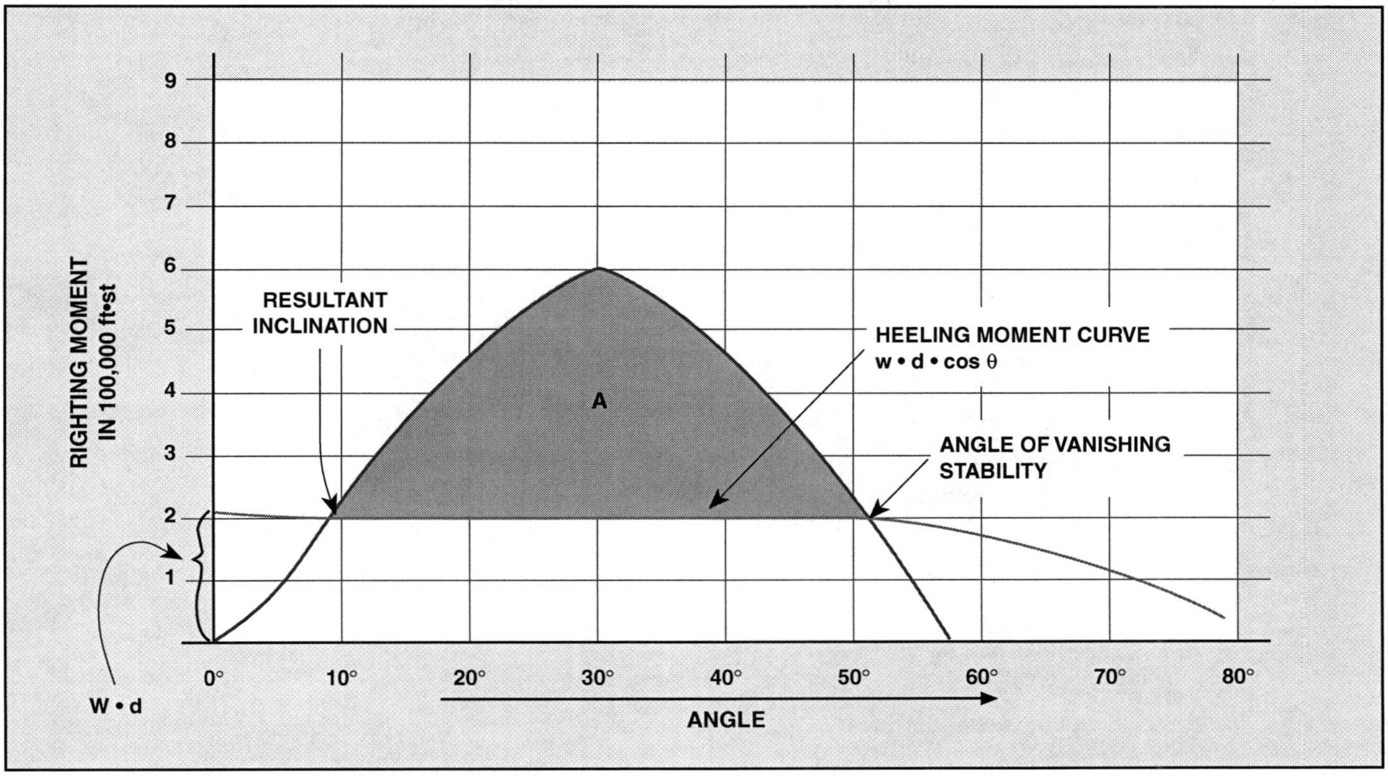

Figure 6.3 Righting moment curve after flooding

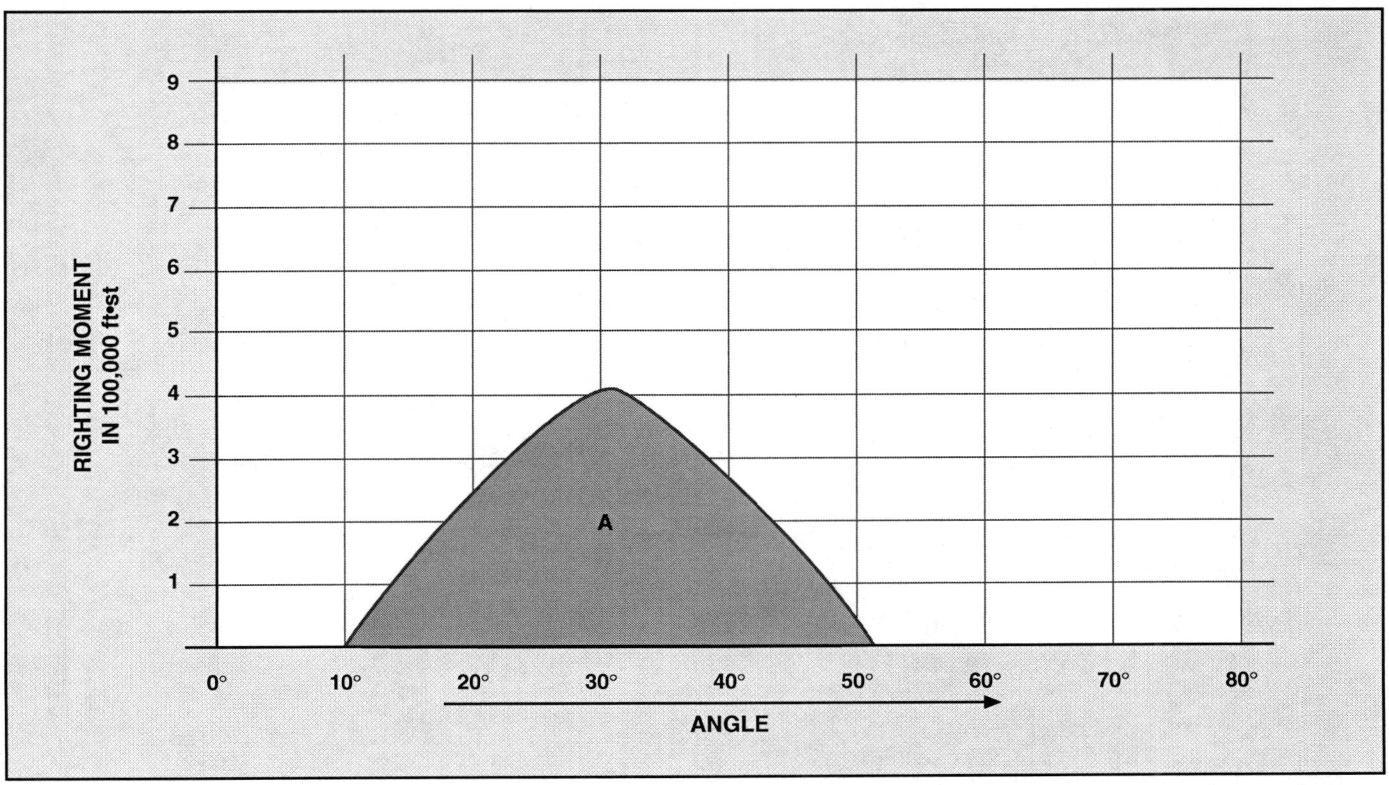

Figure 6.4 Righting moment curve after deduction of heeling moment curve

4. Most important is to maintain the watertight integrity of the rig and minimize additional flooding.
   - ❖ Seal off the damaged compartment.
   - ❖ Minimize additional flooding into adjacent compartments by immediately closing all remote- and local-operated watertight doors, ventilation shafts, and weather-tight closures. If applicable, close all valves of the bilge and ballast system serving the compartment and the adjacent area. Verify all hatches and manholes in the area are closed.
   - ❖ Take remote and manual sounding of the damaged compartment and all adjacent compartments.
   - ❖ Monitor the draft changes at regular intervals.
5. Once the flooding is contained, start the counteraction procedure to return the rig to an acceptable heel and trim condition. This will involve either to counterballast a tank on the high side or to deballast a tank on the low side of the rig. If installed, consider using the secondary deballast system (like on Rig Type 1 series).
6. To counterballast has the advantage because:
   - ❖ It is faster than deballasting because of the higher flow rates.
   - ❖ Ballasting can be accomplished by free flooding through the pumps.
   - ❖ It is possible to deballast a tank after the rig is returned to an acceptable heel and trim.
   - ❖ Even though the mean draft will increase with ballasting, the freeboard at the lowest corner will increase while the rig is returned to upright position.
7. The disadvantage of counterballasting is that the draft of the rig will increase in addition to an increase of draft from the flooding.
8. The option to counterballast is only possible if the wave and swell condition will endanger the underdeck structure by wave impact. Deballasting is then the only option.
9. If a forward tank or compartment is damaged, the main ballast pump may not be able to overcome the suction lift. (The maximum suction lift is 32 ft.) Figure 6.5 shows examples from the rig's marine operations manual of two rigs addressing the possibility of suction problems.
10. At light drafts, not enough tanks may be available with ballast water. Counterballast may then be the only option. In extreme list or heel cases, the ballast intake suction may be above the waterline.
11. If the trim and/or heel after the damage is such that the ballast tank vents are submerged or close to the waterline, close all ballast tank valves from the remote station. The auxiliary vent lines extend above the main deck.

Excerpted from Rig Type 3's marine operations manual—

Estimate of Maximum Trim Angles

Ballast Tank	Full (deg)	1/2 Full (deg)	Empty (deg)	Deballast Ratio
B1	9.0	5.5	1.1	0.78
B2	11.5	7.1	2.5	0.53
B3	10.8	6.5	2.2	0.59
B4	19.5	12.7	5.3	0.18
B5	19.5	12.7	5.3	0.18

The deballast ratio is an indication of the unit's change in trim in proportion to the reduction in the tank's water level. A deballast ratio of less than one indicates that even though pumping could be started, the pump suction will be lost at some point due to the excessive suction lift requirement.

If suction cannot be attained because of excessive lift, open the suction valves for two tanks and take suction

Excerpted from Rig Type 2's marine operations manual—

Estimate of Maximum Trim Angles

Ballast Tank	Full (deg)	1/2 Full (deg)	Empty (deg)	Ratio
B1	7.0	4.1	1.1	2.4
B2/B3	8.9	5.3	1.7	1.25
B4‡	17.1	7.1	2.6	Col 0.12 Hull 0.48
B5	11.5	7.1	2.6	0.56
B6	17.5	11.0	4.5	0.13

‡B4 extends to the 35 ft flat in the 18 ft column.

Figure 6.5 Examples of suction limits

6.13 DAMAGE STABILITY CALCULATION ON BOARD

To be prepared at all times, the barge supervisor should practice, at regular intervals, calculations of assumed damage situations for various anticipated conditions. Examples of damage stability calculations are part of the rig's marine operations manual (MOM). Preprinted standard spreadsheets are used to calculate (with the added weight method) the final result of the flooding and a separate section allows for completing the counteraction to bring the rig back to an even heel and trim condition. In addition, the computer allows quick test cases using the same added weight principle.

Figure 6.6 shows an example of a calculation on damage stability spreadsheet and subsequent counteraction for a Rig Type 2 semisubmersible.

With the information from the actual load condition, hydrostatic properties, and various tank tables, the sequence of the calculations is basically always the same:

1. Start with the information of the initial load condition before the damage—
   - Initial rig drafts on each column.
   - Initial displacement and KG.
2. With information from various tables, find the weight of the water ingress with the corresponding vertical, longitudinal, and transverse moments.
3. Calculate the bodily sinkage with TPI or displacement change.
4. Calculate the new average draft, displacement, and corresponding VCG, GM_T, and GM_L.
5. Apply the FSE to VCG, GM_T, and GM_L.
6. Calculate with MH1° and MT1° the heel and trim. Convert degrees into ft or m.
7. To find the damage condition drafts, add and deduct the heel and trim values to each corner of the column.
8. Decide what will be the best appropriate counteraction for the specific condition.
9. Find the VCG, LCG, and TCG and corresponding moments of the counterweight.
10. Calculate the final average draft, displacement, and corresponding VCG, GM_T, and GM_L.
11. Apply the FSE to VCG, GM_T, and GM_L.
12. Calculate with MH1° and MT1° the corrective heel and trim. Convert degrees into ft or m.
13. To find the corrective final draft, apply the heel and trim values to the damaged condition draft.

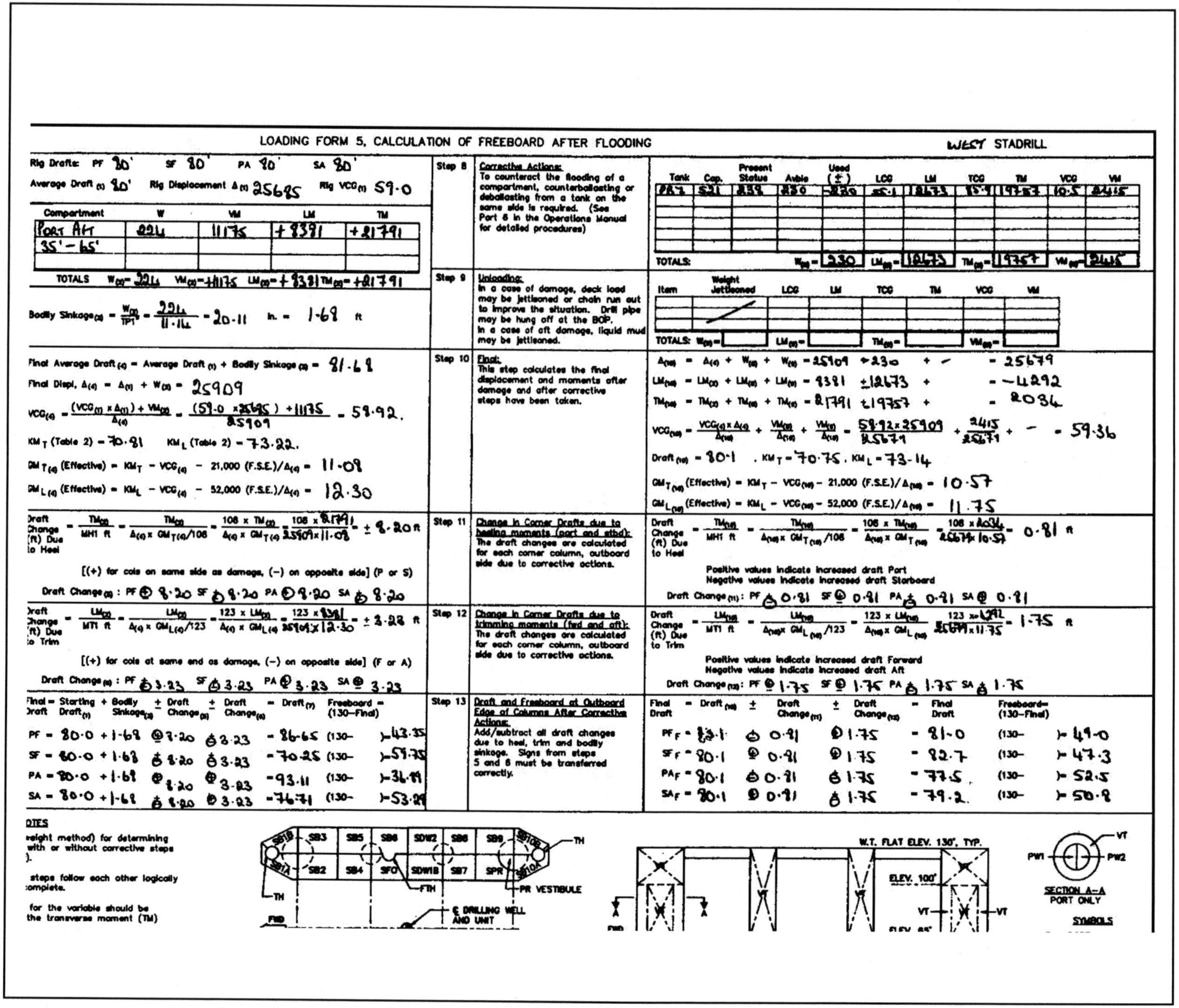

LOADING FORM 5, CALCULATION OF FREEBOARD AFTER FLOODING — WEST STADRILL

Rig Drafts: PF 80' SF 80' PA 80' SA 80'

Average Draft (1) 80' Rig Displacement $\Delta_{(1)}$ 25685 Rig $VCG_{(1)}$ 59.0

Compartment	W	VM	LM	TM
Port Aft 35'–65'	2214	11175	+8391	+21791
TOTALS	$W_{(2)}$ = 2214	$VM_{(2)}$ = +11175	$LM_{(2)}$ = +8391	$TM_{(2)}$ = +21791

Bodily Sinkage(3) = $W_{(2)}$/TPI = 2214/11.16 = 20.11 in. = 1.69 ft

Final Average Draft(4) = Average Draft(1) + Bodily Sinkage(3) = 81.69

Final Displ. $\Delta_{(4)}$ = $\Delta_{(1)}$ + $W_{(2)}$ = 25909

$VCG_{(4)}$ = ($VCG_{(1)}$ x $\Delta_{(1)}$) + $VM_{(2)}$ / $\Delta_{(4)}$ = (59.0 x 25685) + 11175 / 25909 = 59.92

KM_T (Table 2) = 70.81 KM_L (Table 2) = 73.22

$GM_{T(4)}$ (Effective) = KM_T – $VCG_{(4)}$ – 21,000 (F.S.E.)/$\Delta_{(4)}$ = 11.08

$GM_{L(4)}$ (Effective) = KM_L – $VCG_{(4)}$ – 52,000 (F.S.E.)/$\Delta_{(4)}$ = 12.30

Draft Change (ft) Due to Heel = $TM_{(2)}$/MH1 ft = $TM_{(2)}$ / $\Delta_{(4)}$ x $GM_{T(4)}$/108 = 108 x $TM_{(2)}$ / $\Delta_{(4)}$ x $GM_{T(4)}$ = 108 x 21791 / 25909 x 11.08 = ± 8.20 ft

[(+) for cols on same side as damage, (–) on opposite side] (P or S)

Draft Change(5): PF ⊕ 8.20 SF ⊖ 8.20 PA ⊕ 8.20 SA ⊖ 8.20

Draft Change (ft) Due to Trim = $LM_{(2)}$/MT1 ft = $LM_{(2)}$ / $\Delta_{(4)}$ x $GM_{L(4)}$/123 = 123 x $LM_{(2)}$ / $\Delta_{(4)}$ x $GM_{L(4)}$ = 123 x 8391 / 25909 x 12.30 = ± 3.23 ft

[(+) for cols at same end as damage, (–) on opposite side] (F or A)

Draft Change(6): PF ⊖ 3.23 SF ⊖ 3.23 PA ⊕ 3.23 SA ⊕ 3.23

Final Draft	= Starting Draft(1)	+ Bodily Sinkage(3)	± Draft Change(5)	± Draft Change(6)	= Draft(7)	Freeboard = (130–Final)
PF	80.0	+1.69	⊕ 8.20	⊖ 3.23	86.65	(130–)= 43.35
SF	80.0	+1.69	⊖ 8.20	⊖ 3.23	70.25	(130–)= 59.75
PA	80.0	+1.69	⊕ 8.20	⊕ 3.23	93.11	(130–)= 36.89
SA	80.0	+1.69	⊖ 8.20	⊕ 3.23	76.71	(130–)= 53.29

Step 8 — Corrective Actions: To counteract the flooding of a compartment, counterballasting or deballasting from a tank on the same side is required. (See Part 6 in the Operations Manual for detailed procedures)

Tank	Cap.	Present Status	Avble	Used (±)	LCG	LM	TCG	TM	VCG	VM
PB2	521	239	230	–230	55.1	12673	85.9	19757	10.5	2415
TOTALS:				$W_{(8)}$ = 230		$LM_{(8)}$ = 12673		$TM_{(8)}$ = 19757		$VM_{(8)}$ = 2415

Step 9 — Unloading: In a case of damage, deck load may be jettisoned or chain run out to improve the situation. Drill pipe may be hung off at the BOP. In a case of aft damage, liquid mud may be jettisoned.

Item	Weight Jettisoned	LCG	LM	TCG	TM	VCG	VM
TOTALS:	$W_{(9)}$ =		$LM_{(9)}$ =		$TM_{(9)}$ =		$VM_{(9)}$ =

Step 10 — Final: This step calculates the final displacement and moments after damage and after corrective steps have been taken.

$\Delta_{(10)}$ = $\Delta_{(4)}$ + $W_{(8)}$ + $W_{(9)}$ = 25909 + 230 + — = 25679

$LM_{(10)}$ = $LM_{(2)}$ + $LM_{(8)}$ + $LM_{(9)}$ = 8391 ± 12673 + = –4292

$TM_{(10)}$ = $TM_{(2)}$ + $TM_{(8)}$ + $TM_{(9)}$ = 21791 ± 19757 + = 2034

$VCG_{(10)}$ = $VCG_{(4)}$ x $\Delta_{(4)}$/$\Delta_{(10)}$ + $VM_{(8)}$/$\Delta_{(10)}$ + $VM_{(9)}$/$\Delta_{(10)}$ = 59.92 x 25909 / 25679 + 2415 / 25679 + — = 59.36

Draft(10) = 80.1, KM_T = 70.75, KM_L = 73.14

$GM_{T(10)}$ (Effective) = KM_T – $VCG_{(10)}$ – 21,000 (F.S.E.)/$\Delta_{(10)}$ = 10.57

$GM_{L(10)}$ (Effective) = KM_L – $VCG_{(10)}$ – 52,000 (F.S.E.)/$\Delta_{(10)}$ = 11.75

Step 11 — Change in Corner Drafts due to heeling moments (port and stbd): The draft changes are calculated for each corner column, outboard side due to corrective actions.

Draft Change (ft) Due to Heel = $TM_{(10)}$/MH1 ft = $TM_{(10)}$ / $\Delta_{(10)}$ x $GM_{T(10)}$/108 = 108 x $TM_{(10)}$ / $\Delta_{(10)}$ x $GM_{T(10)}$ = 108 x 2034 / 25679 x 10.57 = 0.81 ft

Positive values indicate increased draft Port. Negative values indicate increased draft Starboard.

Draft Change(11): PF ⊖ 0.81 SF ⊕ 0.81 PA ⊖ 0.81 SA ⊕ 0.81

Step 12 — Change in Corner Drafts due to trimming moments (fwd and aft): The draft changes are calculated for each corner column, outboard side due to corrective actions.

Draft Change (ft) Due to Trim = $LM_{(10)}$/MT1 ft = $LM_{(10)}$ / $\Delta_{(10)}$ x $GM_{L(10)}$/123 = 123 x $LM_{(10)}$ / $\Delta_{(10)}$ x $GM_{L(10)}$ = 123 x 4292 / 25679 x 11.75 = 1.75 ft

Positive values indicate increased draft Forward. Negative values indicate increased draft Aft.

Draft Change(12): PF ⊕ 1.75 SF ⊕ 1.75 PA ⊖ 1.75 SA ⊖ 1.75

Step 13 — Draft and Freeboard at Outboard Edge of Columns After Corrective Actions: Add/subtract all draft changes due to heel, trim and bodily sinkage. Signs from steps 5 and 6 must be transferred correctly.

Final Draft	= Draft(10)	± Draft Change(11)	± Draft Change(12)	= Final Draft	Freeboard = (130–Final)
PF_F	83.1	⊖ 0.81	⊕ 1.75	81.0	(130–)= 49.0
SF_F	80.1	⊕ 0.81	⊕ 1.75	82.7	(130–)= 47.3
PA_F	80.1	⊖ 0.81	⊖ 1.75	77.5	(130–)= 52.5
SA_F	80.1	⊕ 0.81	⊖ 1.75	79.2	(130–)= 50.8

NOTES

weight method) for determining with or without corrective steps).

steps follow each other logically complete.

for the variable should be the transverse moment (TM)

Figure 6.6 Example damage stability calculation

EXERCISE
Chapter 6

Calculate at least one out of three questions.

1. Use the following hydrostatic properties and tables.
   Rig Type 2A in deep tow condition accidentally floods a previously empty PDW2 to half depth with fresh water.
   VCG of the tank is 20 ft aft of LCF and 108 ft to port of centerline.
   Towing draft = 55 ft
   Initial KG = 62 ft
   LCF is at the centerline of the rig.
   Length between draft marks is 225 ft.
   Width between draft marks is 195 ft.
   Calculate the final draft of each column.
   Calculate final GM_{wet}, KG longitudinal, and KG transverse.

2. Rig Type 2A is WOW with a deck load of 2,200 st at a draft of 69.9 ft. VCG = 56.2 ft. Damage occurs to port aft 30 ft diameter column between 35 ft and 65 ft flats. The permeability of the flooded compartment is 95%. Calculate the final draft for each column and the new KG with a compartment permeability of 95%.
   Displacement at 69.95 ft = 25,958 st.
   Length between draft marks = 225 ft. Width = 195 ft.
   Center of the flooded column is 97.5 ft from LCF.
   Center of the flooded column is 112.5 ft from TCF.
   COF is at the centerlines.

3. Rig Type 2A is loaded as shown on the attached load form summary page. Calculate the final drafts for each column if the port aft column is damaged and floods between the 51-ft and 65-ft horizontal flats.

Table 2
RIG TYPE 2A
HYDROSTATIC PROPERTIES
(Page 2 of 4)

DRAFT ft	DISPL ST	KB ft	TP1" ST	KML ft	KMT ft	MH1deg (ft•st) VCG=50	VCG=60	VCG=70	MT1deg (ft•st) VCG=50	VCG=60	VCG=70
21.1	18480	10.56	11.69	91.67	83.01	10646	7421	4196	13439	10214	6989
22.0	18605	10.63	11.69	91.20	82.59	10582	7335	4088	13378	10131	6884
23.0	18745	10.72	11.69	90.69	82.15	10518	7246	3975	13312	10040	6769
24.0	18890	10.82	13.08	90.18	81.70	10450	7153	3857	13245	9948	6651
25.0	19055	10.93	13.66	89.60	81.19	10374	7048	3723	13169	9843	6517
26.0	19215	11.06	13.08	89.08	80.74	10309	6955	3602	13106	9753	6399
27.0	19360	11.17	11.69	88.61	80.33	10248	6869	3490	13045	9667	6288
28.0	19500	11.29	11.69	88.17	79.95	10193	6789	3386	12990	9587	6184
29.0	19640	11.41	11.33	83.74	77.50	9426	5998	2571	11565	8137	4710
30.0	19775	11.53	11.33	83.36	77.17	9377	5926	2475	11513	8062	4611
31.0	19910	11.66	11.33	83.00	76.85	9330	5855	2380	11467	7992	4517
32.0	20045	11.80	11.67	82.66	76.55	9289	5791	2293	11427	7928	4430
33.0	20255	12.01	21.77	82.14	76.09	9224	5689	2154	11361	7826	4291
34.0	20550	12.32	26.75	81.44	75.48	9139	5553	1966	11277	7690	4104
35.0	20880	12.67	28.21	81.23	74.80	9039	5395	1751	11381	7737	4093
36.0	21215	13.03	26.73	81.03	74.15	8943	5240	1538	11490	7787	4085
37.0	21505	13.35	21.74	80.44	73.65	8876	5123	1370	11423	7670	3917
38.0	21715	13.58	11.60	80.02	73.30	8829	5039	1249	11376	7586	3796
39.0	21855	13.74	11.60	79.75	73.07	8799	4985	1171	11347	7533	3719
40.0	21995	13.90	11.60	79.49	72.86	8775	4937	1098	11320	7482	3643
41.0	22135	14.07	11.60	79.25	72.66	8754	4891	1028	11300	7436	3573
42.0	22270	14.24	11.60	79.02	72.46	8729	4843	956	11279	7392	3506
						VCG=50	VCG=60	VCG=65	VCG=50	VCG=60	VCG=65
43.0	22410	14.42	10.62	69.28	67.32	6774	2863	907	7541	3629	1674
44.0	22535	14.58	10.62	69.14	67.18	6757	2824	857	7528	3595	1628
45.0	22665	14.75	10.62	69.00	67.06	6748	2793	815	7516	3560	1582
46.0	22790	14.92	10.64	68.88	67.00	6762	2784	795	7509	3532	1543
47.0	22920	15.10	10.73	68.82	67.26	6904	2904	904	7528	3528	1528
48.0	23050	15.28	10.84	68.78	67.58	7072	3049	1038	7555	3532	1521
49.0	23180	15.47	10.88	68.70	67.64	7136	3091	1068	7565	3520	1497
50.0	23310	15.66	10.88	68.59	67.52	7127	3059	1025	7563	3495	1460
51.0	23440	15.85	10.88	68.49	67.41	7122	3031	986	7564	3473	1428
52.0	23570	16.05	10.88	68.40	67.30	7116	3003	946	7569	3455	1399
53.0	23700	16.25	10.88	68.31	67.21	7118	2982	914	7573	3437	1369
54.0	23830	16.45	10.89	68.24	67.12	7120	2961	882	7586	3427	1347
55.0	23965	16.66	10.97	68.70	67.31	7240	3057	966	7821	3639	1548
56.0	24095	16.87	11.08	69.32	67.60	7401	3196	1093	8124	3919	1817
57.0	24230	17.09	11.14	69.65	67.73	7498	3269	1154	8309	4081	1966
58.0	24365	17.31	11.14	69.58	67.65	7505	3253	1127	8326	4074	1948
59.0	24495	17.54	11.14	69.52	67.57	7511	3236	1099	8345	4070	1932
60.0	24630	17.77	11.14	69.47	67.49	7518	3220	1070	8369	4071	1921

Note:

1. All properties listed are based upon actual displacement. Displacement accounts for additions to molded hull such as plate thicknesses, anchor racks, thrusters and fairleaders. See Figures 20, 21 and 22 for explanation of symbols.

Table 2
RIG TYPE 2A
HYDROSTATIC PROPERTIES
(Page 3 of 4)

DRAFT ft	DISPL ST	KB ft	TP1" ST	KML ft	KMT ft	MH1deg (ft-ST) VCG=50	MH1deg (ft-ST) VCG=60	MH1deg (ft-ST) VCG=65	MT1deg (ft-ST) VCG=50	MT1deg (ft-ST) VCG=60	MT1deg (ft-ST) VCG=65
61.0	24765	18.00	11.14	69.42	67.43	7533	3211	1050	8393	4071	1910
62.0	24900	18.23	11.14	69.38	67.37	7548	3203	1030	8422	4076	1903
63.0	25030	18.47	11.14	69.34	67.31	7562	3193	1009	8448	4080	1896
64.0	25165	18.71	11.14	69.31	67.27	7585	3193	997	8481	4089	1893
65.0	25300	18.95	11.14	69.29	67.24	7612	3197	989	8517	4102	1894
66.0	25430	19.19	11.14	69.27	67.20	7634	3195	976	8552	4114	1895
67.0	25565	19.44	11.14	69.25	67.17	7661	3199	968	8589	4127	1896
68.0	25700	19.69	11.14	69.24	67.15	7692	3207	964	8630	4144	1902
69.0	25835	19.94	11.14	69.24	67.13	7724	3215	960	8675	4166	1912
70.0	25965	20.20	11.14	69.24	67.12	7758	3226	961	8719	4187	1921
70.5	26035	20.33	11.14	69.24	67.11	7774	3231	959	8742	4198	1927
71.0	26100	20.46	11.14	69.25	67.11	7794	3239	961	8769	4213	1936
71.5	26170	20.58	11.14	69.25	67.11	7815	3247	964	8792	4225	1941
72.0	26235	20.72	11.14	69.26	67.11	7834	3255	966	8818	4240	1950
72.5	26300	20.85	11.14	69.27	67.11	7853	3263	968	8845	4255	1960
73.0	26370	20.98	11.14	69.28	67.11	7874	3272	971	8873	4271	1970
73.5	26435	21.11	11.14	69.29	67.12	7898	3285	978	8900	4286	1979
74.0	26500	21.24	11.14	69.30	67.12	7918	3293	980	8926	4301	1989
74.5	26570	21.38	11.14	69.31	67.13	7943	3306	988	8954	4317	1999
75.0	26635	21.51	11.14	69.32	67.14	7967	3319	995	8981	4332	2008
75.5	26705	21.65	11.14	69.34	67.15	7993	3332	1002	9014	4353	2023
76.0	26770	21.78	11.14	69.35	67.16	8017	3345	1009	9040	4368	2032
76.5	26835	21.91	11.14	69.37	67.17	8041	3358	1016	9072	4388	2047
77.0	26905	22.05	11.14	69.39	67.18	8067	3371	1024	9105	4409	2061
77.5	26970	22.19	11.14	69.41	67.20	8096	3389	1036	9136	4429	2076
78.0	27035	22.33	11.14	69.43	67.21	8120	3402	1043	9168	4449	2090
78.5	27105	22.46	11.14	69.45	67.23	8151	3420	1055	9201	4470	2105
79.0	27170	22.60	11.14	69.47	67.25	8180	3438	1067	9232	4491	2120
79.5	27235	22.74	11.14	69.50	67.27	8209	3456	1079	9269	4516	2139
80.0	27305	22.88	11.14	69.52	67.28	8235	3469	1087	9302	4537	2154
80.5	27370	23.02	11.14	69.55	67.31	8269	3492	1103	9338	4562	2173
81.0	27440	23.16	11.14	69.58	67.33	8299	3510	1116	9377	4588	2193
81.5	27505	23.30	11.14	69.60	67.35	8328	3528	1128	9409	4608	2208
82.0	27570	23.44	11.14	69.63	67.38	8363	3551	1145	9445	4634	2228
82.5	27640	23.59	11.14	69.66	67.40	8393	3570	1158	9484	4660	2248
83.0	27705	23.73	11.14	69.70	67.43	8428	3593	1175	9525	4690	2273
83.5	27770	23.87	11.14	69.73	67.46	8462	3616	1192	9562	4716	2292
84.0	27840	24.02	11.14	69.76	67.49	8498	3639	1210	9601	4742	2313
84.5	27905	24.16	11.14	69.80	67.52	8532	3662	1227	9643	4773	2338
85.0	27975	24.31	11.14	69.83	67.55	8568	3686	1245	9682	4799	2358

Table 2
RIG TYPE 2A
HYDROSTATIC PROPERTIES
(Page 4 of 4)

DRAFT	DISPL	KB	TP1"	KML	KMT	MH1deg (ft-ST)			MT1deg (ft-ST)		
ft	ST	ft	ST	ft	ft	VCG=50	VCG=60	VCG=65	VCG=50	VCG=60	VCG=65
85.5	28040	24.45	11.14	69.87	67.59	8608	3714	1267	9724	4830	2383
86.0	28105	24.60	11.14	69.91	67.62	8643	3738	1285	9766	4861	2408
86.5	28175	24.74	11.14	69.95	67.65	8679	3762	1303	9810	4893	2434
87.0	28240	24.89	11.14	69.99	67.69	8719	3790	1326	9852	4924	2459
87.5	28305	25.04	11.14	70.03	67.73	8758	3819	1349	9895	4955	2485
88.0	28375	25.18	11.14	70.07	67.77	8800	3848	1372	9939	4987	2511
88.5	28440	25.33	11.14	70.11	67.81	8840	3876	1395	9982	5018	2536
89.0	28510	25.48	11.14	70.15	67.85	8882	3906	1418	10026	5050	2562
89.5	28575	25.63	11.14	70.20	67.89	8922	3935	1441	10074	5087	2593
90.0	28640	25.78	11.14	70.25	67.93	8962	3964	1465	10122	5123	2624
90.5	28710	25.93	11.14	70.29	67.97	9004	3993	1488	10166	5156	2651
91.0	28775	26.08	11.14	70.34	68.02	9050	4028	1517	10215	5193	2682
91.5	28840	26.23	11.14	70.39	68.06	9090	4057	1540	10263	5230	2713
92.0	28910	26.38	11.14	70.44	68.11	9137	4092	1569	10313	5267	2745
92.5	28975	26.53	11.14	70.49	68.16	9183	4126	1598	10361	5305	2776
93.0	29040	26.69	11.14	70.54	68.21	9229	4161	1627	10410	5342	2808
93.5	29110	26.84	11.14	70.59	68.26	9277	4196	1656	10461	5380	2840
94.0	29175	27.00	11.14	70.64	68.31	9323	4231	1685	10509	5418	2872
94.5	29245	27.15	11.14	70.70	68.36	9371	4267	1715	10565	5461	2909
95.0	29310	27.30	11.14	70.75	68.41	9417	4302	1744	10614	5499	2941

Note:

1. All properties listed are based upon actual displacement. Displacement accounts for additions to molded hull such as plate thicknesses, anchor racks, thrusters and fairleaders. See Figures 20, 21 and 22 for explanation of symbols.

RIG TYPE 2A

FREE SURFACE EFFECT OF HULL TANK LIQUIDS

Sounding (FT)	PB1A SB10A	SB1A PB10A	PB1B SB10B	SB1B PB10B	B2,B3 B6,B9	B4,B5 B7,B8	DW1 DW2	FO
			FSE (ft-ST) about Longitudinal Axis (Roll)					
0	0	0	0	0	0	0	0	0
1	290	777	227	227	970	847	826	818
3	499	1098	421	421	1428	1247	1216	1204
6	690	1380	596	596	1850	1615	1576	1560
9	776	1505	677	677	2050	1791	1747	1729
10.5	783	1515	682	1454	2074	1812	1767	1749
12	784	784	683	1456	2050	1791	1747	1729
15	744	744	646	1404	1850	1615	1576	1560
18	633	633	543	1251	1428	1247	1216	1204
20	477	477	398	1037	970	847	826	818
21	0	0	0	0	0	0	0	0
			FSE (ft-ST) about Transverse Axis (Pitch)					
0	0	0	0	0	0	0	0	0
1	974	1615	630	630	3522	2355	2297	2971
3	1443	2011	1144	1144	4013	2684	2618	3385
6	1630	2195	1336	1336	4378	2928	2856	3693
9	1672	2245	1345	1345	4533	3032	2957	3823
10.5	1681	2254	1354	2025	4550	3043	2968	3837
12	1678	1678	1384	2022	4533	3032	2957	3823
15	1666	1666	1375	2002	4378	2928	2856	3693
18	1605	1605	1321	1922	4013	2684	2618	3385
20	1526	1526	1258	1809	3522	2355	2297	2971
21	0	0	0	0	0	0	0	0
			FSE (ft-ST) about Diagonal Axis 49 Deg from Longitudinal					
0	0	0	0	0	0	0	0	0
1	680	1254	457	457	2424	1706	1664	2044
3	1037	1618	833	833	2900	2066	2015	2446
6	1225	1844	1017	1017	3290	2363	2305	2775
9	1286	1926	1057	1057	3464	2498	2436	2922
10.5	1294	1936	1065	1779	3484	2513	2451	2938
12	1293	1293	1082	1778	3464	2498	2436	2922
15	1269	1269	1061	1745	3290	2363	2305	2775
18	1187	1187	986	1633	2900	2066	2015	2446
20	1074	1074	888	1477	2424	1706	1664	2044
21	0	0	0	0	0	0	0	0

Notes:

1. To obtain the reduction in allowable VCG due to excess free surfaces, enter the tabulated values for each tank containing liquid on Form 3. Each value is the free surface effect for one tank.
2. For moderate weather conditions use roll FSE for the draft range 18 to 72 ft and use diagonal axis FSE for the draft range 72 tc 95 ft.
3. For storm conditions use roll FSE for the draft range 55 to 61 ft and use diagonal axis FSE for the draft range above 61 ft.

RIG TYPE 2A
WEIGHT AND MOMENT EFFECT OF FLOODED COMPARTMENTS

Compartment	Perm (%)	Weight (ST)	Vertical Moment VCG (ft)	WxVCG (ft-ST)	LCG (ft)	Overturning Moments WxLCG (ft-ST)	TCG (ft)	WxTCG (ft-ST)
Pump Room w/o Vestibule	94%	514	10.5	5394	- 89.5	-45997	+/- 83.7	+/-42992
Vestibule	98%	109	10.5	1139	-106.0	-11499	+/- 91.0	+/- 9871
Ballast Tank PB1A	95%	350	10.8	3763	+126.0	+44125	+ 86.7	+30366
Ballast Tank SB1A	95%	404	10.0	4048	+127.4	+51466	- 88.2	-35621
Ballast Tank PB1B	95%	335	10.8	3605	+125.3	+41974	+108.7	+36412
Ballast Tank SB1B	95%	388	11.5	4447	+126.8	+49237	-107.1	-41609
Ballast Tank B2	95%	629	10.5	6602	+ 92.4	+58107	+/- 85.0	+/-53422
Ballast Tank B3 or B9	95%	629	10.5	6602	+/- 92.4	+/-58107	+/-110.0	+/-69192
Ballast Tank B4 or B7	95%	550	10.5	5773	+/- 55.1	+/-30289	+/- 85.0	+/-46706
Ballast Tank B5 or B8	95%	550	10.5	5773	+/- 55.1	+/-30289	+/-110.1	+/-60506
Ballast Tank B6	95%	629	10.5	6602	+ 17.4	+10947	+/-110.0	+/-69192
Ballast Tank PB10A	95%	404	10.0	4048	-127.4	-51466	+ 88.2	+35621
Ballast Tank SB10A	95%	350	10.8	3763	-126.0	-44125	- 86.7	-30366
Ballast Tank PB10B	95%	388	11.5	4447	-126.8	-49237	+107.1	+41609
Ballast Tank SB10B	95%	335	10.8	3605	-125.3	-41974	-108.7	-36412
Fuel Oil Tank FO	95%	629	10.5	6602	+ 17.4	+10947	+/- 85.0	+/-53422
Drill Water Tank DW1	95%	550	10.5	5773	- 19.9	-10947	+/- 85.0	+/-46706
Drill Water Tank DW2	95%	550	10.5	5773	- 19.9	-10947	+/-110.1	+/-60506
Columns 21 to 35 (f&a)	98%	253	28.0	7070	+/-112.5	+/-28406	+/- 97.5	+/-24619
Columns 35 to 51 (f&a)	98%	282	43.1	12171	+/-112.5	+/-31770	+/- 97.5	+/-27534
Columns 51 to 65 (f&a)	98%	260	58.1	15118	+/-112.5	+/-29273	+/- 97.5	+/-25370
Col Annulus Outbd (f&a)	98%	104	82.5	8564	+/-120.7	+/-12530	+/-105.7	+/-10973
Col Annulus Inbd (f&a)	98%	185	82.5	15246	+/-115.1	+/-21269	+/- 88.5	+/-16353
Col Annulus Mdshp (f&a)	98%	151	82.5	12449	+/-103.4	+/-15603	+/-102.4	+/-15445
Col Annulus Centr (f&a)	98%	235	82.5	19379	+/-112.5	+/-26426	+/- 97.5	+/-22903
Columns 100 to 130(f&a)	98%	638	115.0	73417	+/-112.5	+/-71821	+/- 97.5	+/-62245
Chain Lkr Nos. 1,4,5,8	98%	92	61.4	5627	+/-112.5	+/-10316	+/-106.5	+/- 9766
Chain Lkr Nos. 2,3,6,7	98%	92	62.4	5731	+/-121.5	+/-11166	+/- 97.5	+/- 8960
18 ft Column 21 to 35	98%	97	21.0	2037	+/- 37.5	+/- 3638	+/- 97.5	+/- 9458
18 ft Column 35 to 65	98%	222	50.0	11105	+/- 37.5	+/- 8329	+/- 97.5	+/-21655
18 ft Column 65 to 130	98%	519	97.5	50573	+/- 37.5	+/-19451	+/- 97.5	+/-50573
Elevator Shaft (f&a)	99%	95	75.5	7165	+/-105.0	+/- 9965	+/- 92.5	+/- 8781
Truss @ 30 ft Col (f&a)	98%	176	35.0	6160	+/-112.5	+/-19800	0.0	0
Truss @ 18 ft Col (f&a)	98%	176	35.0	6160	+/- 37.5	+/- 6600	0.0	0

Note:

1. Ballast tank capacities in the tank capacity tables are slightly higher than the ones shown above because their capacities are based upon 98% permeability instead of 95%.

SEDCO 712 LOAD FORM 13 FEB 86 PG1

This is a PLANNED load case.

OPERATOR______________________ WATER DEPTH = 700.00 ft PDW2-F

LOCATION______________________ FUEL OIL S.G. = .84 SDW2-F

________LAT________LONG ANCHOR WEIGHT = 23.50 st

WATCHSTANDER__________________

SUPERVISOR____________________ Values in ( ) are for MEASURED drafts

VESSEL STATUS	VCG ALW	VCG MRGN	DKLD MRGN	SEDCO 712 VESSEL LOADING 13 FEB 86 13:21		CORNER CALC	DRAFTS (MEAS)
STORM	Not Allowed	---)		DISP=27773. st (0.)	DKLD=2600. st	SF= 83.5	(0.)
OPERATE	59.2	.0	1.	DRAFT= 83.5 ft (0.0)	VCG =59.23 ft	PF= 83.5	(0.)
60'STORM	63.2	-2.3	-450.	TRIM = -.0 dg (0.00)	LCG = -.00 ft	SA= 83.5	(0.)
70'STORM	56.2	-6.4	-1337.	HEEL = .0 dg (0.00)	TCG = .00 ft	PA= 83.5	(0.)

PG	PAGE 14 SUMMARY PAGE	WEIGHT tons	VM(t-ft) +UP	VCG(ft) +UP	LM(t-ft) +FWD	LCG(ft) +FWD	TM(t-ft) +PORT	TCG(ft) +PORT
1	P/S PIPE DECK	350.0	53008.	151.45	-26995.	-77.13	4515.	12.90
2	PORT/STBD DECKS	129.5	17363.	134.07	-6700.	-51.74	-1663.	-12.84
3	DRY BULK STORAGE	590.0	50910.	86.29	-3660.	-6.20	4160.	7.05
4	SACK STORAGE	90.0	12060.	134.00	-6660.	-74.00	-3830.	-42.56
5	MUD HOUSE & MISC	131.3	18124.	138.04	-3817.	-29.07	-821.	-6.25
6	DRILL AREA LOADS	217.1	34799.	160.29	197.	.91	-992.	-4.57
7	SETBACK LOADS	35.0	7210.	206.00	0.	0.00	0.	0.00
8	MUD PITS	503.1	67030.	133.23	-34935.	-69.44	-2057.	-4.09
9	MISC TANKS	333.4	33508.	100.52	26472.	79.41	24346.	73.03
10	VOID TANKS	0.0	0.	0.00	0.	0.00	0.	0.00
11	PORT TANKS	5179.6	48576.	9.38	-7727.	-1.49	509198.	98.31
12	STARBOARD TANKS	5700.9	54526.	9.56	49532.	8.69	-553978.	-97.17
13	ANCHOR+RISR+LSHP	14512.9	1247873.	85.98	14292.	.98	21121.	1.46
PG	TOTAL WT & MOM	27772.8	1644987.	59.23	-0.	-.00	0.	.00

P/S PIPE DECK	WEIGHT	VCG	LCG	TCG	PAGE 1	WEIGHT	VCG	LCG	TCG
5' DRILL PIPE	131.0	153	-75	0		0.0	0	0	0
HEAVY WT DP	21.0	153	-60	40		0.0	0	0	0
8" DC	0.0	153	-90	10		0.0	0	0	0
SLIP JOINTS	32.0	151	-75	0		0.0	0	0	0
MISC DRILL TOOLS	10.0	151	-75	0		0.0	0	0	0
9.5"DC(5 JNTS)	85.0	150	-90	25		0.0	0	0	0
6.5 DC(14 JNTS)	51.0	150	-60	30		0.0	0	0	0
PUP/TSIJT/PEND	10.0	151	-95	42		0.0	0	0	0
SUBS/STAB/GR/HK	10.0	150	-110	-40		0.0	0	0	0
	0.0	0	0	0		0.0	0	0	0
	0.0	0	0	0		0.0	0	0	0
	0.0	0	0	0		0.0	0	0	0
	0.0	0	0	0		0.0	0	0	0
TOTAL WT & MOM	WGT=	350.0	VM=	53008.0	LM=	-26995.0	TM=	4515.0	

PORT/STBD DECKS	WEIGHT	VCG	LCG	TCG	PAGE 2	WEIGHT	VCG	LCG	TCG
PRODUCTION EQUIP	20.0	138	-80	75		0.0	0	0	0
	0.0	0	0	0		0.0	0	0	0
DIVING	16.0	138	20	-20		0.0	0	0	0
RISER	50.0	132	-40	0		0.0	0	0	0
1 BAKER TANK	25.0	135	-80	-70		0.0	0	0	0
MISC OTIS EQUIP	2.0	132	-10	-65		0.0	0	0	0
OTIS LAB	1.0	135	-10	75		0.0	0	0	0
SPIDER /MISC	13.0	131	-105	-75		0.0	0	0	0
DOPE &GREASE	2.5	127	-10	-25		0.0	0	0	0
	0.0	0	0	0		0.0	0	0	0
	0.0	0	0	0		0.0	0	0	0
	0.0	0	0	0		0.0	0	0	0
	0.0	0	0	0		0.0	0	0	0
TOTAL WT & MOM	WGT=	129.5	VM=	17362.5	LM=	-6700.0	TM=	-1662.5	

DRY BULK STORAGE	WEIGHT	VCG	LCG	TCG	PAGE 3	WEIGHT	VCG	LCG	TCG
BULK TK NO 1 P/F	150.0	87	38	99		0.0	0	0	0
BULK TK NO 1 S/F	105.0	89	38	-99		0.0	0	0	0
BULK TK NO 2 P/A	80.0	90	-38	99		0.0	0	0	0
BULK TK NO 3 P/A	80.0	75	-38	99		0.0	0	0	0
BULK TK NO 2 S/A	80.0	90	-38	-99		0.0	0	0	0
BULK TK NO 3 S/A	80.0	75	-38	-99		0.0	0	0	0
MUD SURGE TK NO1	5.0	141	-74	-17		0.0	0	0	0
MUD SURGE TK NO2	5.0	141	-78	-22		0.0	0	0	0
CEMENT SURGE TK	5.0	141	-86	-20		0.0	0	0	0
	0.0	0	0	0		0.0	0	0	0
	0.0	0	0	0		0.0	0	0	0
	0.0	0	0	0		0.0	0	0	0
	0.0	0	0	0		0.0	0	0	0
TOTAL WT & MOM	WGT=	590.0	VM=	50910.0	LM=	-3660.0	TM=	4160.0	

Drilling, 2600 ST Deck Load, 83.5 ft Draft, Moderate Seas (3) 1 of 3

CHAPTER 7
The Ballast System

7.1 INTRODUCTION

Ballast systems on board vessels were developed to enable a vessel to float at a specific draft required for transit. In addition, forward and aft trim tanks would give the vessel the possibility to use the most efficient trim condition for transit. A subdivision of a double bottom provided for the necessary amount of tanks, each connected with pipelines and valves to pumps to fill or empty the tanks.

With the arrival of jackups, submersibles, and semisubmersibles in the offshore drilling world a sophisticated ballast system was developed over the years to interchange the draft of the mobile offshore units from transit draft to operations draft. Jackup ballast systems are mainly used to trim the rig during transit and to preload the jackup to test the soil strength before the start of the drilling operation.

In this chapter, mainly the ballast systems used on semisubmersibles are used as an example to explain the principles of ballast systems.

7.2 THE MAIN BALLAST SYSTEM COMPONENTS

The main components of a ballast system on a semisubmersible are:

1. *The ballast tanks*. Each pontoon is divided identically by watertight bulkheads into a set of tanks. A specific amount of tanks are designated as ballast tanks. The remaining tanks may be for drill water or fuel tanks. Figures 7.1 and 7.2 are typical tank divisions for Rig Type 1 and Rig Type 2.

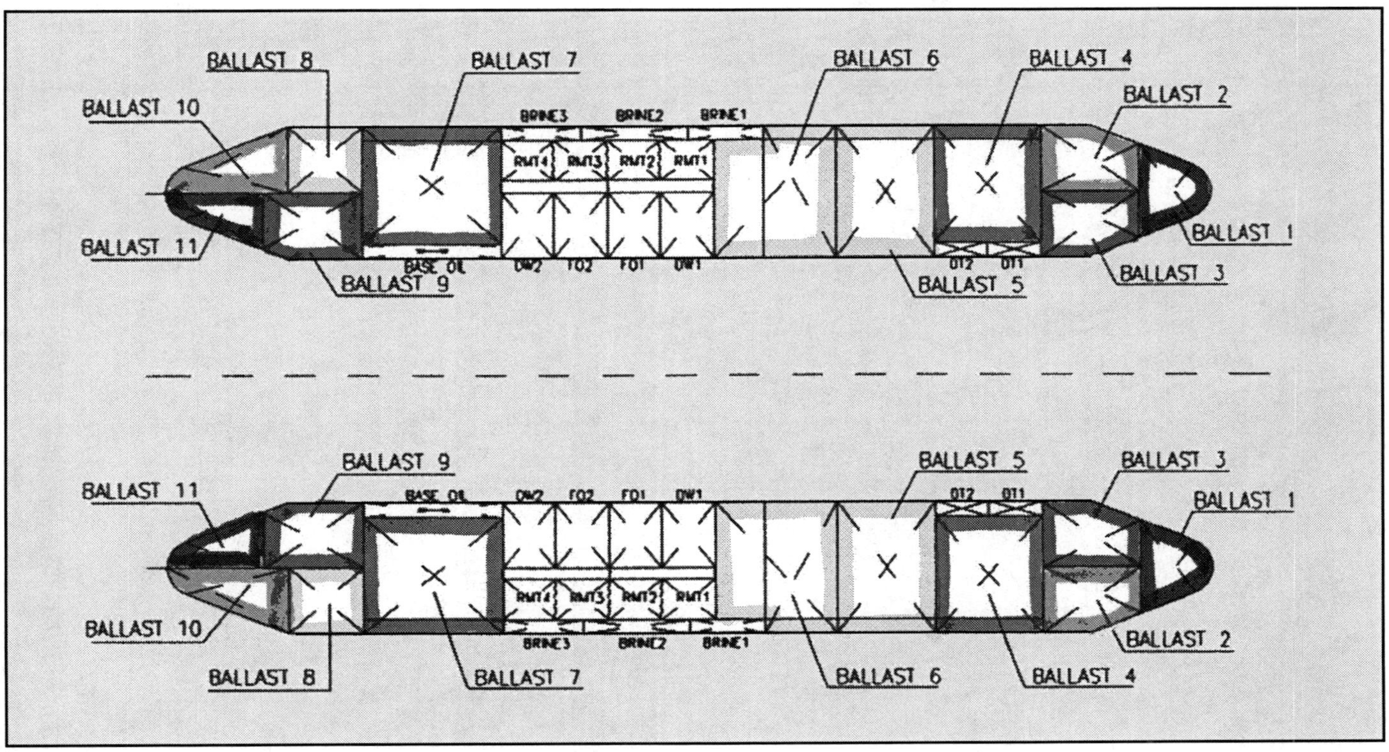

Figure 7.1 Example tank division Rig Type 1

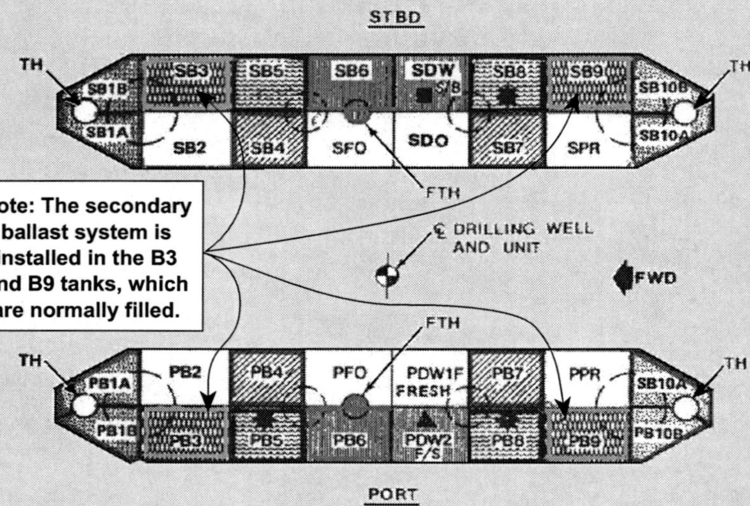

SYMBOLS

- P PORT
- S STARBOARD
- B BALLAST TANK
- FO FUEL OIL TANK
- DW DRILL WATER TANK
- TH PROPULSION THRUSTER
- FTH FUTURE THRUSTER
- DO DRILL OIL TANK

SETS OF TANKS

	SET	TOTAL BALLAST WEIGHT (SHORT TONS OF SEA WATER)	AVERAGE NET CHANGE OF DRAFT ON COLUMNS (FEET)
TRIM TANKS			
BALLAST SET A	A	2460	19
BALLAST SET B	B	2080	16
BALLAST SET C	C	2300	17.5
BALLAST SET D	D	2400	18

✸ INCLUDES VOLUME IN 18 FT. COLUMNS UP TO ELEV' +35'-0"
▲ SEA WATER OR FRESH WATER (COMBINATION TANK) F/S
■ SEA WATER OR BRINE (COMBINATION TANK) S/B

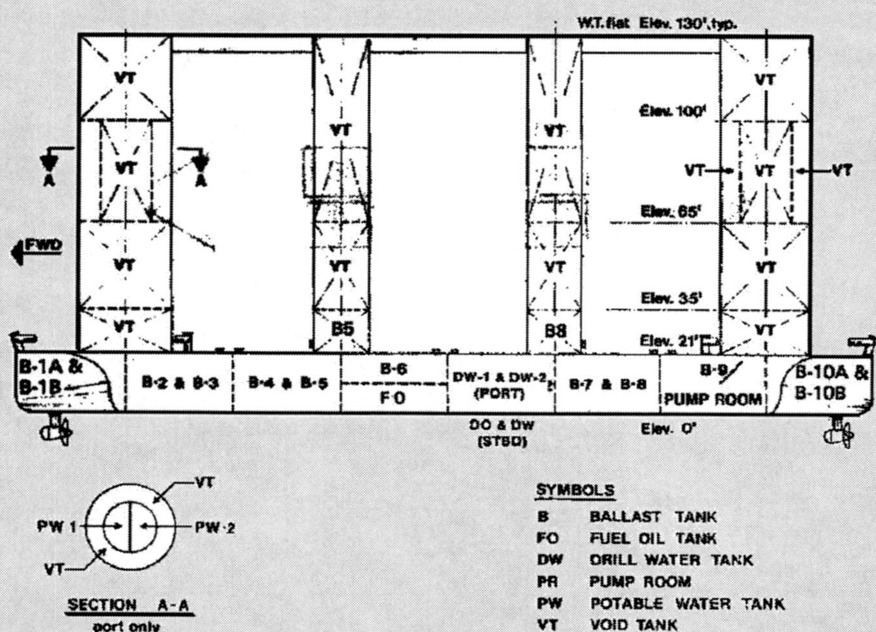

SYMBOLS

- B BALLAST TANK
- FO FUEL OIL TANK
- DW DRILL WATER TANK
- PR PUMP ROOM
- PW POTABLE WATER TANK
- VT VOID TANK

Figure 7.2 Example tank arrangement Rig Type 2

2. *The ballast piping system.* The pipelines provide the connection between the tanks and the pumps. Interconnection between forward and aft sections. Of the pontoons and even cross connections between pontoons may be possible (figs. 7.3 and 7.4).
3. *Pumps.* Electrically driven centrifugal pumps provide means to fill or discharge into each ballast tank (fig. 7.5).
4. *Valves.* Numerous valves including check valves are installed to control the supply or discharge of the fluid in the piping system. Most valves are remote-controlled with the possibility of local manual control. All remote-controlled valves are fitted with a fail-safe close function activated in case of power loss (figs. 7.6 and 7.7).
5. *Sea chests.* Each pump is connected to a sea chest. Intake of ballast water is admitted through a sea chest with a strainer and sea suction header to the pump (fig. 7.8).
6. *Overboard discharge.* Independent discharge lines are connected to each pump. The overboard discharge lines discharge at a high level above the waterline.
7. *Vent lines.* All ballast tanks have vent lines which terminate through the columns via watertight closures. Watertight closures below the waterline are fitted with a remote-control device. Vent lines to the main deck terminate with a gooseneck and ball closure.
8. *Tank sounding system.* Each tank is fitted with sensors to read the sounding from a control room. In addition, manual soundings are possible, or a system in the pump room provides for additional remote sounding readings.

7.3 THE BALLAST SYSTEM

The ballast system is remotely controlled from a main console on the bridge or ballast control room.

Depending on the type of rig and layout, two or four pump rooms are located in the pontoons. Generally, at least four ballast pumps are installed. Each pump is equipped with electrically-driven centrifugal pumps. The pump capacity as an example can be 2,500 gpm with a head of 90 ft or 650 m^3/hr @ 3 bar.

Ballast is admitted through a sea chest, strainer, and sea suction header (fig. 7.8) to the pump and then to the ballast tank header with lines to each tank. Ballast is discharged by opening the ballast header pump suction valves from a tank and the overboard discharge lines. Each pump serves a section of tanks in the particular hull.

As shown in the diagrams in Figures 7.3 and 7.4, a normally closed manual or remote-controlled crossover or block valve separates the forward and aft sections. The function of this block valve is:

1. To isolate the forward and aft sections to prevent equalization between forward and aft tanks in case of improperly closed or leaking valves. On some rigs like the Rig Type 1 series, a crossover line between the pontoons allows cross transfer between the pontoons. The block valve in the crossover line must remain closed in normal operations for the same reason as the forward-aft block valve.
2. To have redundancy in case of pump failure between tank sections forward and aft and, if applicable, between port and starboard pontoons.
3. To enable the transfer of ballast from forward to aft and, if applicable, from port to starboard, or vice versa.

As mentioned, in addition to the piping system between forward and aft tanks, ballast systems may have crossover lines between pontoons which gives even more possibilities in case of failure of one section of the ballast system.

On many rigs a crossover line to the bilge system is installed so the ballast may act as emergency bilge pumps.

The remote control to the actuator (fig. 7.7) can be hydraulic, pneumatic, or electric. Remote-controlled valves and particularly the sea chest valves are protected with a fail-safe close device in case of power loss. This will prevent free flooding or equalizing tank levels.

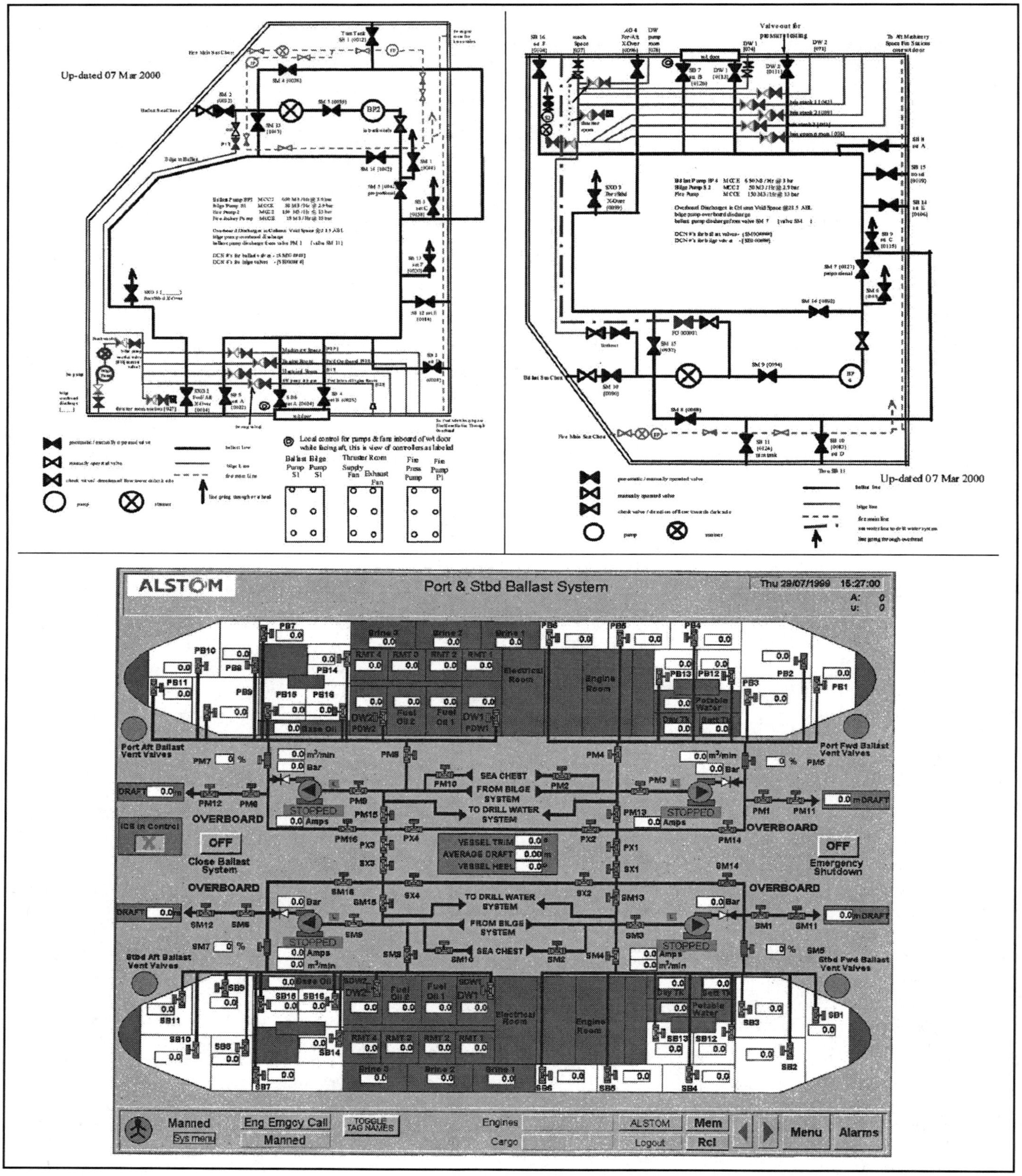

Figure 7.3 Example of computer schematic of the ballast piping system and details of schematic of the SB forward and aft pump rooms of Rig Type 1.

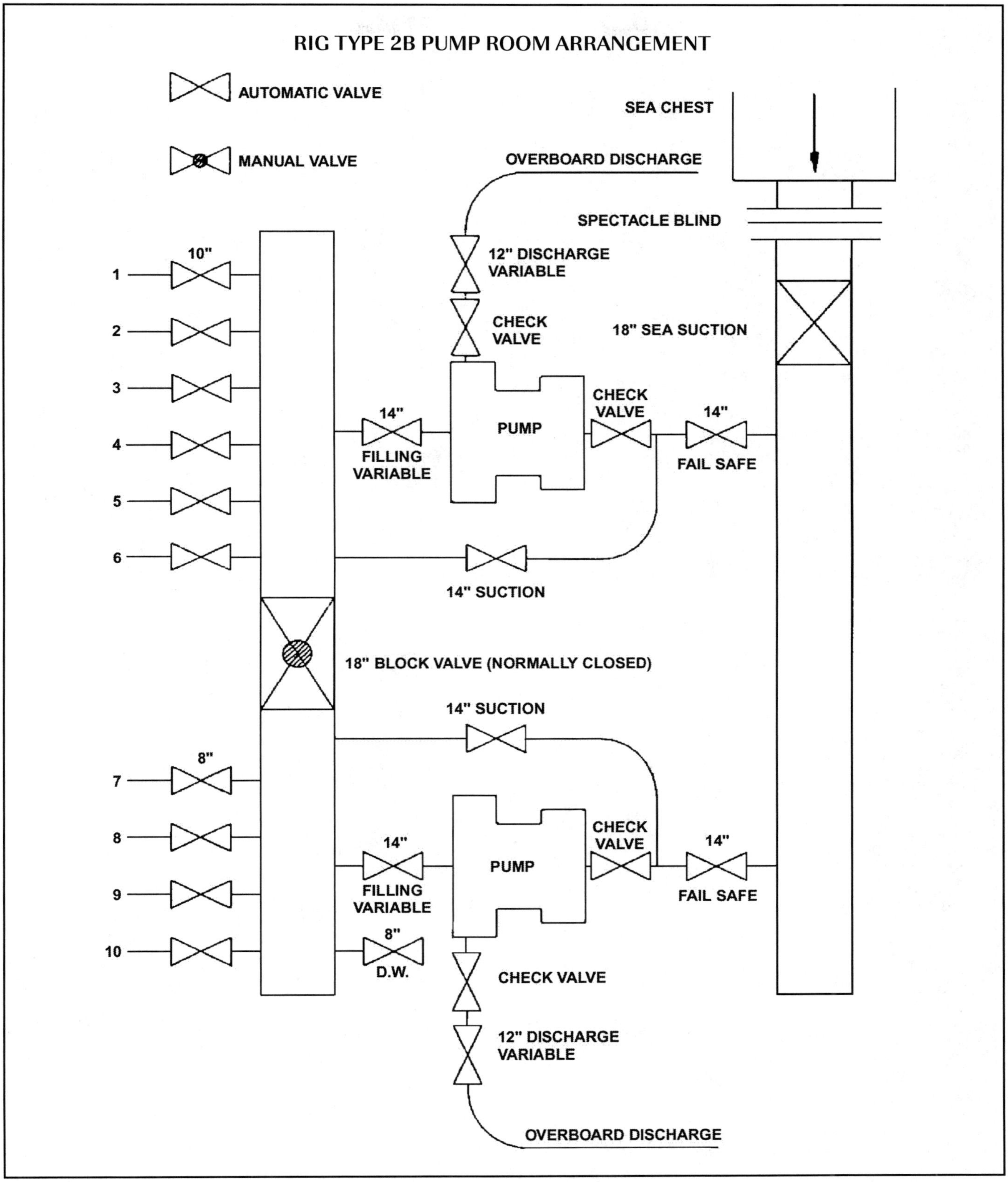

Figure 7.4 Example schematic piping ballast system in pump room of Rig Type 2 series

Figure 7.5 Example of ballast pump and fire-fighting pump set up in forward pump room on Rig Type 1

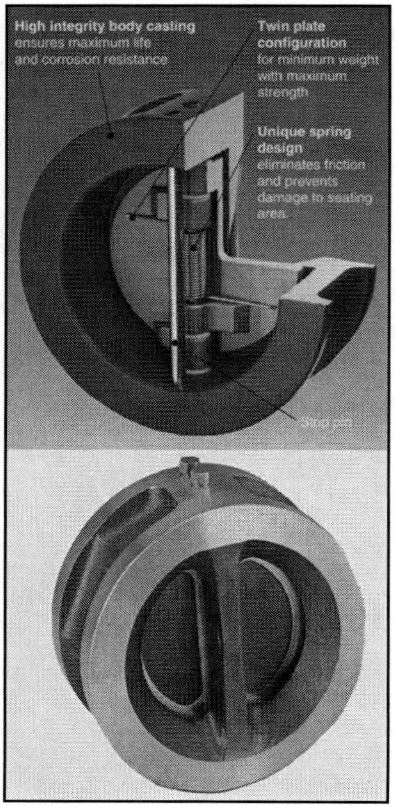

Figure 7.6 Example of a double-sided check valve

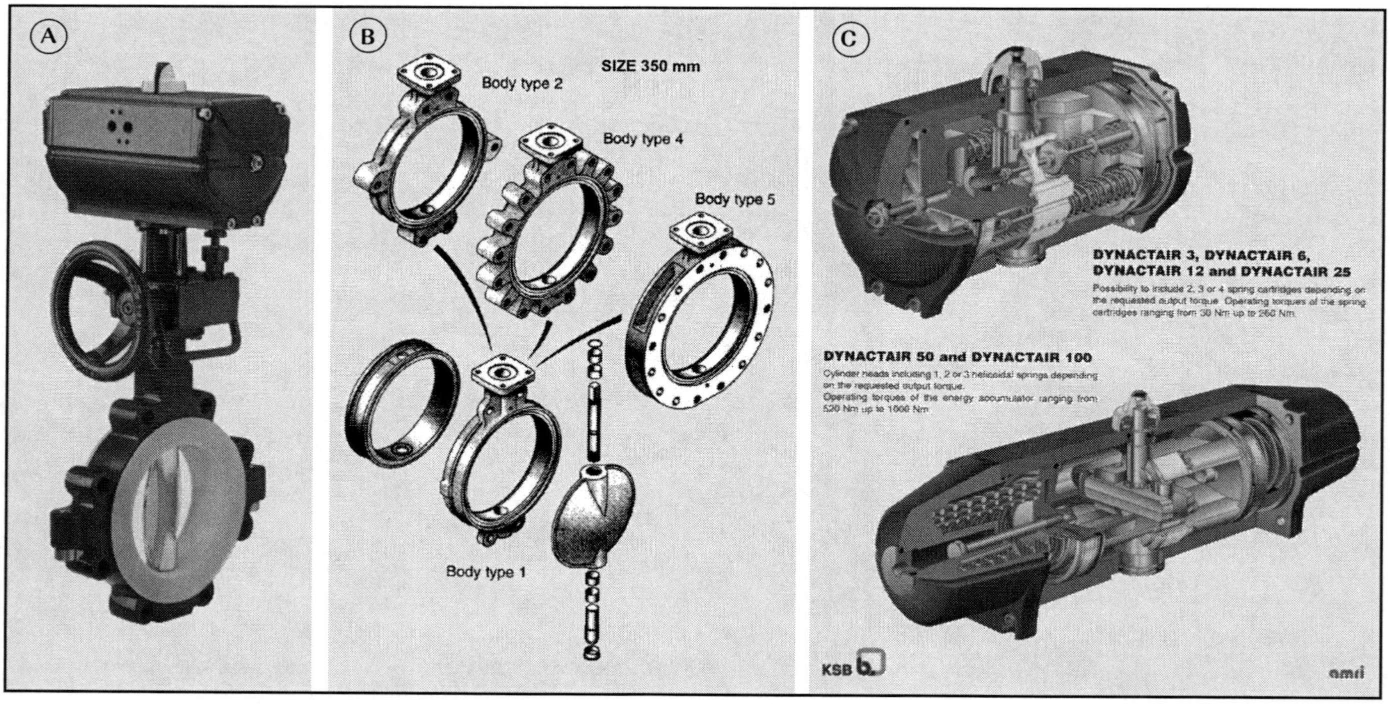

Figure 7.7 A. Example of remote-controlled butterfly valve with local manual override. B. Explosion view of butterfly valve parts. C. Remote-controlled actuator with fail-safe close device.

The valves can be butterfly valves (see fig. 7.7), gate valves, or ball valves. A remote tank-gauging system is normally installed in the ballast control room. Manual sounding through sounding pipes to each tank provides a backup. Various types of remote tank-reading systems, such as electronic sensing systems based on absolute pressure or purge air bubble systems, are used in the industry.

Figure 7.8 Example of a sea chest with manual and remotely-operated valves on Rig Type 1

7.4 THE SECONDARY DEBALLAST SYSTEM

On some rigs an independent secondary deballast system (SDS) is installed (fig. 7.9). A typical example is the system used on the Rig Type 2 series. The system consists of:

1. Four independent electrically-driven 1,200 gpm submerged deep well

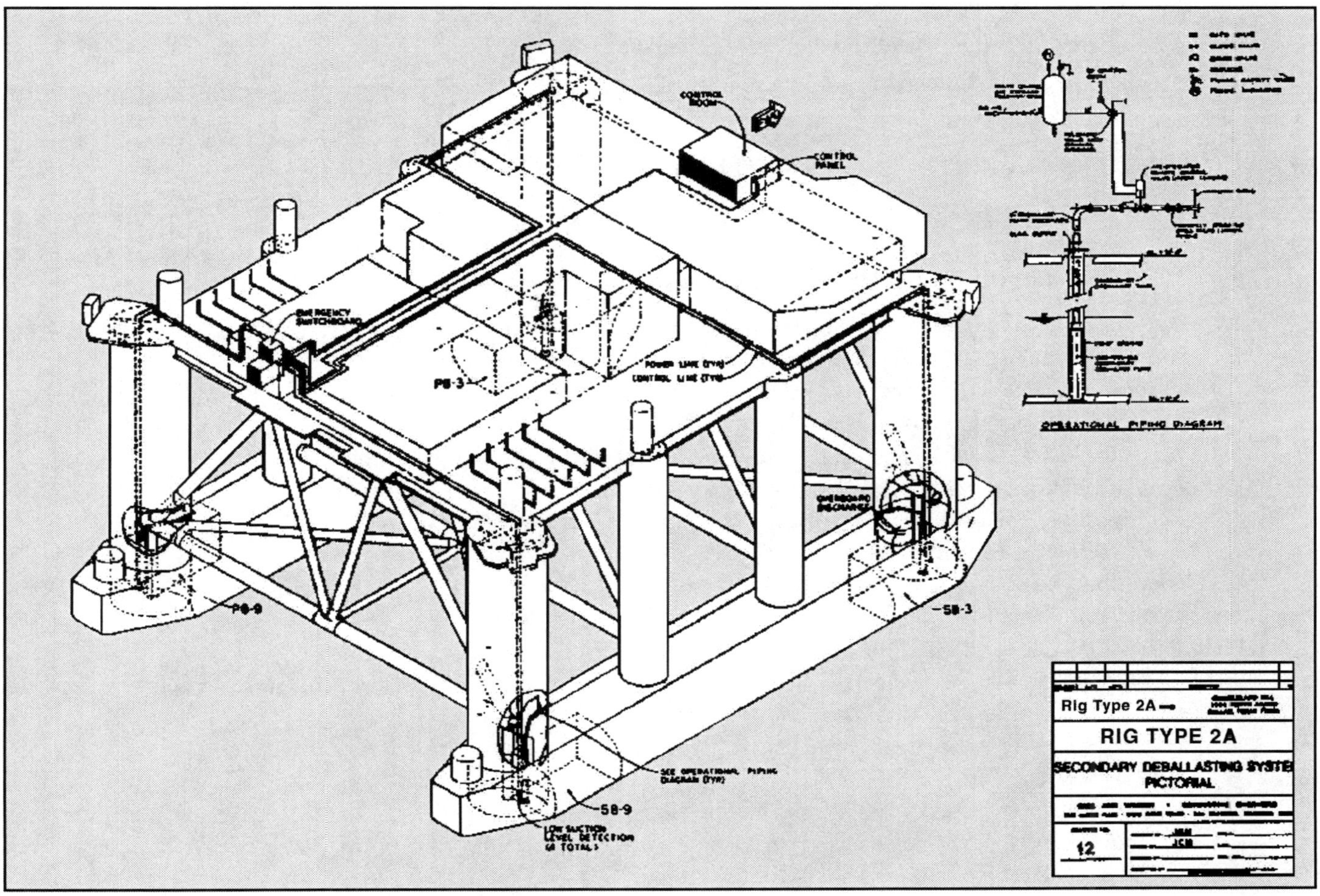

Figure 7.9 Picture with secondary deballast system (SDS)

pumps with associated valves and overboard discharge lines.

2. Four ballast tanks which are normally filled. For the Rig Type 2 series the tanks are the B3 and B9 tanks (see fig. 7.2).
3. A remote-controlled console in the barge control room and an emergency switchboard with associated control lines.
4. Connection to main power and emergency power. (Only one pump at the time)
5. The SDS is designed only to deballast.

The system has important advantages, such as:

- Operation is possible separately or in conjunction with the main ballast system.
- Independent operation with main power or emergency power (with one pump).
- In case of main pump room failure by flooding or fire, the SDS system can be used to counteract flooding by deballasting.
- In case of failure of the ballast system, the SDS system can deballast to survival draft.
- In case of large trim caused by a major compartment flooding, the SDS system can counteract by deballasting a large trim which may make the ballast pump nonoperational.

7.5 THE MAIN FUNCTIONS OF THE BALLAST SYSTEM

The three main functions of the ballast system are:

1. To ballast and deballast the drilling rig between light draft, operations draft, and survival draft.
2. To maintain the rig's heel and trim at zero degrees.
3. To counteract accidental flooding or damaged flooding.

7.6 GENERAL REMARKS ON BALLASTING AND DEBALLASTING PROCEDURES

The following general remarks on ballasting and deballasting are guidance notes based on generally accepted procedures. The rig specific procedures may be different as operating procedures and ballast systems vary from rig to rig. Always follow the rig specific instructions and procedures as explained in the marine operations manual (MOM).

1. Only qualified trained personnel should operate the ballast system.
2. Never leave the ballast control system unattended while operating the system.
3. Before starting the ballast system, verify that all systems are 100% operational.
4. Verify the initial draft and required final draft.
5. Calculate the initial and final stability including heel and trim. Verify that the allowable VCG remains within the limits throughout the process.
6. Confirm deck load and deck load margins.
7. Adjust anchor chain tensions as required.
8. To prevent equalizing of tank water levels, always keep all valves closed when not in the process of ballasting/deballasting.
9. Maintain even heel and keel during the process of ballasting/deballasting.
10. Ballast tank sets as specified in the rig's marine operation manual (MOM). Asymmetrical loading will impose excessive stress on the rig structure.
11. In deballast mode, avoid running the ballast pumps dry for an extended period of time.
12. Avoid slack tanks.
13. Do not take any workboats alongside while in the process of ballasting/deballasting. At certain drafts, an outer shell may not protect the columns, or damage to a horizontal flat will cause a two-compartment flooding.
14. Fill ballast tanks as close as possible to the intake level.
15. After completion of the ballast/deballast operation, verify that all valves are returned to the closed position.

16. Finally strip all empty tanks. A slight heel or trim for each tank may be required to allow the suction pipe to draw water until the tanks are completely empty.
17. The rig will experience somewhat increased motions when the tubulars start to submerge. This is normal. The motions will decrease after the tubulars are completely submerged. Do not try to compensate for the motions.

7.7 BALLASTING

The following procedures are general and should be used as a guidance to start and stop ballasting. Rig specific procedures may be different and should be followed as explained in the MOM. The following procedure is based on filling the tanks with the ballast pumps:

- Fill the ballast tanks with the pumps. (Gravity fill is used too, but this is rig specific and follows a different procedure.)
- Verify that the variable valves at the pressure side of the pumps are closed.
- Verify that all other valves are closed.
- Open the sea chest (sea inlet valve) followed by the tank valve to be filled and the valves in the pressure line from the pump.
- Start the appropriate ballast pump(s).
- Open the variable (regulating) valve until the pump reads the required pressure.
- To finish the ballast operation, it is important to prevent slamming of the check valve:
    1. Close the ballast pump discharge valve.
    2. Immediately shut down the pump after confirmation that the discharge valve is closed.
    3. Close the sea chest valve.
    4. Close the tank manifold valve.
- After completion of ballasting, always verify all valves are closed.

7.8 DEBALLASTING

The following procedures are general and should be used as a guidance to start and stop deballasting. Rig specific procedures may be different and should be followed as explained in each rig's marine operation manual:

- Verify that the sea chest valve, pump suction valve, and pump discharge valve to the header are all closed.
- Open the tank suction and tank manifold valve.
- Start the appropriate ballast pump(s).
- Open the variable valve until the pump reads the required pressure.
- To prevent cavitation of the impeller, trim or heel the rig slightly to the suction side of an almost empty tank.
- To finish the deballast operation it is important to prevent slamming of the check valve:
    1. Close the overboard discharge valve.
    2. Shut down the pump immediately after confirmation that the overboard discharge valve is closed.
    3. Close the tank manifold valves and other valves.
- After completion of deballasting, always verify all valves are closed.

Note 1: The effect of slamming for the deballast operation is more dangerous because:

- The additional head of the high level of the overboard discharge.
- The opposite flow between discharge and head.

Slamming can break the parts of the check valve flaps. The broken parts will go through the pump and cause severe damage.

Note 2: Cavitation is the effect of air bubbles in the fluid. The vacuum of the air bubbles around the impeller in the fluid causes water evaporation and pitting in the impeller material.

EXERCISE
Chapter 7

1. Name the main purposes of the ballast system.

2. Name the main components of the ballast system.

3. Why is it necessary to strip almost empty tanks?

4. Explain the sequence to finish the ballast operation.

5. Why is the block valve installed?

6. How can the check valve be damaged?

7. Why are a manual and remote-controlled valve installed close together at the sea chest location (see fig 7.8)?

8. The SDS on the Rig Type 2 series can be used for ballasting and deballasting. True or false?

9. Why is the correct sequence of the shutdown procedure for deballasting even more important than for ballasting?

10. Explain why, in general, it is not allowed to take a workboat alongside during the ballast/deballast operation.

11. Name the advantages of the SDS system.

12. What causes the change in motion when the tubulars start to immerse during ballasting?

13. Explain why it is important to have a fail-safe device installed at the sea chest valve.

14. What is the main reason for cavitation of the impellers?

CHAPTER 8
Design and Construction

8.1 INTRODUCTION

The design of the primary structure of the offshore units involves the sizing, geometrical arrangement, and selection of material for the structural members and their connections. This means that the structure and connections have acceptable fatigue lives and that the allowable stresses are not exceeded when subjected to the maximum loading the unit is expected to withstand.

The units are designed in accordance with accepted practices specified in detail by international regulations such as IMO, Classification Societies, and government authorities.

The rigs are not overdesigned; therefore, the restrictions pertaining to structural safety prescribed in the rig's marine operations manual must be respected.

8.2 FORCES WORKING ON THE FLOATING UNIT

The structure of an offshore drilling rig in operation or under tow is subjected to many forces during its lifetime. These forces or loads can be statical or dynamic.

1. *Statical loads* are loads such as the weight of the rig and its components, the buoyancy force, and the pressure by water.
2. *Dynamic loads* are loads caused by the environmental forces and the rig movement such as roll, pitch, yaw, sway, heave, and surge.

The various combinations of these loads will tend to bend and twist the rig structure. The bending and twisting loads are transferred to the structural members of the rig, which are essential to the overall integrity of the unit.

8.3 STRESS, STRAIN, AND YIELD

An axial pulling load applied to a steel sample produces stress within the material. If the axial pulling progresses, the steel sample starts to elongate. Stress is expressed in load per area; i.e., psi or kg/cm^3. The elongation of the specimen represents the strain in the material. Strain is expressed in in./in. or cm/cm.

The proportional relationship of stress to strain continues until a certain point, and from thereon is not proportional anymore. This is called the *proportional elastic limit* of the material. Before this point, the specimen returns to its original size if the force is removed. After this point, the specimen is permanently deformed.

Continuation of the axial load beyond the point of proportional elastic limit shows that the sample material will elongate without almost any additional load. This is the *yield point* or *tensile yield strength* of the steel sample.

Any additional pulling results in further elongation until the pulling reaches maximum value and then decreases. The material reaches the *ultimate tensile strength* and shortly afterwards the *fracture point*. The various points are displayed in the graph of Figure 8.1.

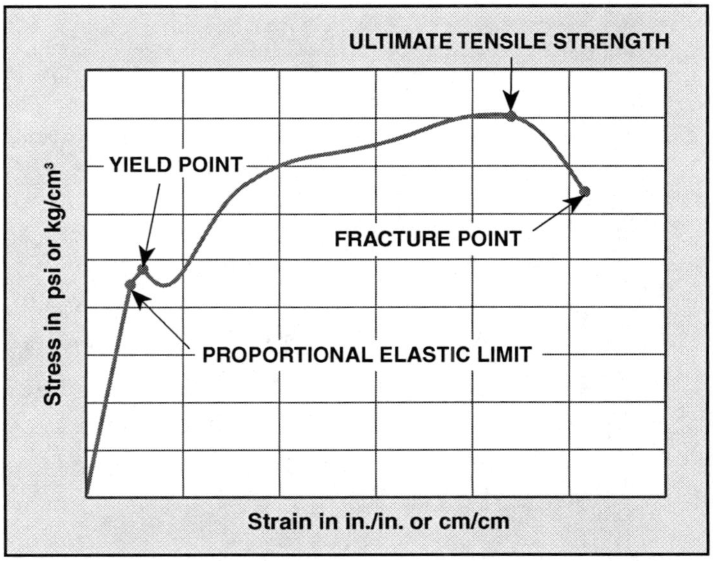

Figure 8.1 Example of a stress-strain diagram

8.4 FATIGUE

Fatigue is the tendency to fracture under cyclic stress. Under high load and increased cycles the fatigue mode may reduce the effective strength of the material. The *endurance limit* is the maximum stress that can be applied to a structure for an indefinite service of life.

8.5 SHEAR STRESS

Shear stress is the actual stress required to shear the material. Shear loads are produced by transverse loads in combination with other loads. Structural failure by surpassing the ultimate shear strength of the structure is very rare.

8.6 SAGGING AND HOGGING

Sagging and hogging are bending forces caused by waves passing by the rig structure or hull. When the wave crest is amidships, the bending force is called hogging. When the wave trough is amidships, the bending force is called sagging (fig. 8.2).

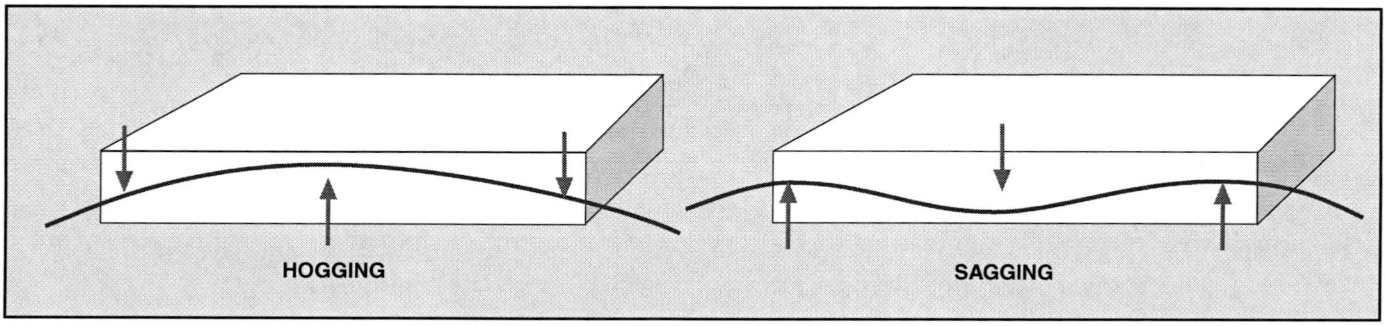

Figure 8.2 Hogging and sagging

8.7 DESIGN LOADING CONDITIONS

The drilling units are designed to withstand various combinations of loads in the operation and transit modes.

1. *Vertical loads.* This includes lightship weight plus all variable loads.
2. *Half-gravity loads* (0.5 G loads). These are loads used to calculate the approximate forces caused by pitch and roll. The combination of vertical loads and half-gravity loads controls the design of the lower hulls, columns, vertical bracings, and the main deck girders.
3. *Transverse loads.* These are the loads that tend to force the structure apart (also called prying loads) and are caused by beam waves with a critical length creating opposing horizontal loads on the lower hull and columns. For a semisubmersible in transit mode with light draft, the transverse load can be considerably high. This type of transversal load is not applicable for the structure of a drilling vessel.
4. *Diagonal loads.* The diagonal loads have the same effect as the transverse loads. These loads are caused by a wave with a crest line running between two opposite corner columns. This creates opposing horizontal forces on the corner columns and causes racking and twisting of the structure. A drilling vessel will only be exposed to a minimum of this type of load.
5. *Diagonal twisting wave loads.* The twisting effect on the structure is caused by the vertical buoyancy change when a wave has a simultaneous crest at two diagonal corner columns and a trough at the two other diagonal corner columns.

Note: Asymmetrical ballasting of the hull tanks on a semisubmersible worsens the effects of the diagonal loads as explained in (4) and (5). Only in emergency damage stability conditions is asymmetric loading allowed. The rig's marine operations manual provides more details on maximum allowable asymmetrical loading conditions. Figure 8.3 is an example of the rig's marine operations manual from the Rig Type 1 showing the importance to minimize asymmetrical loading.

RIG TYPE 1

Marine Operations Manual

PART 3
OPERATING PROCEDURES
ISSUE 1 - OCTOBER 1999

3.1.1.3 **Cross-Hull Twisting Limitations**

A. **The rig's structure and tubulars can be overstressed when the pontoon and column tanks are loaded in such a way to create an excessive difference in the port and starboard longitudinal moments. This difference is called cross-hull twisting. To avoid overstressing the rig's structure and tubulars, the difference in the longitudinal moments shall not exceed those given in the following table:**

Condition	Weather Condition	Cross-Hull Twisting Moment Limitations (Difference Between P & S Longitudinal Moments)
Drilling	Moderate	150 MN•m (55,000 ST•ft)
Transit at shallow & deep & drafts	Moderate	150 MN•m (55,000 ST•ft)
Survival	Storm	70 MN•m (26,000 ST•ft)

NOTES:
The cross-hull twisting moment in the survival condition takes into account the environmental conditions acting on the hull; therefore the limiting moment in this condition is lower than the limiting moment when the rig is drilling or transiting in moderate weather.

Figure 8.3 Example from Rig Type 1's marine operations manual on twisting limitations

6. *Transverse and diagonal horizontal (prying) loads.* These loads are used to compute the stress ranges per foot or metre wave height so that the fatigue life of critical members such as horizontal tubulars and various levels can be checked.

7. *Longitudinal sagging and hogging wave loads.* These loads occur mainly at towing drafts and subject to the lower hulls on semisubmersibles and to vessel-shaped drilling rigs. The sagging loads are maximum when a wave trough is amidships and the hogging forces are maximum when the wave crest is amidships.

8.8 STEEL QUALITY

The Classification Societies prescribe in detail the steel quality and welding procedures for the various sections of the structure. The appropriate steel grades are selected and based on the need for strength, toughness, and ductility in the location used. Figures 8.4 and 8.5 are examples from ABS for ordinary hull structures and higher strength hull structures to show the amount of details used to control the quality of steel to be used.

8.9 MEMBER LOADING AND STRESS

For design purposes, materials and dimensions of structural components are selected to keep total combined stresses below the critical levels.

A *utilization factor* indicates how total combined stresses in a member compares to maximum allowable stresses. For example, under certain design load conditions, the utilization factor for a 6 ft horizontal tubular may be 98%, meaning that these members are stressed to 98% of their maximum allowable stress. Additional loading caused by failure to follow the rig's marine operations manual (MOM) procedures could lead to overstressing of the rig structure. This could cause sudden structural failure, permanent deformation, and accelerated metal fatigue shortening the useful life of the rig.

8.10 FATIGUE IN MEMBERS AND CONNECTIONS

The fatigue tendency is greater at those areas of members and connections that have combinations of the following:

1. High levels of stress change as cyclic loads, such as waves, are applied.
2. Points of high stress concentrated in relatively small areas called stress raisers; i.e., undercut at the toe of welds.
3. Structural discontinuities, so-called hot spots, that affect the load path in a heavily loaded member; i.e., cable-way penetrations in girders.
4. Complex intersections of loaded members; i.e., K joints.
5. Seawater corrosion.

Typical areas on MODUs that are more prone to fatigue failure are:

- Brace connections at column shells
- Vent holes
- Drains and hydrophone wells
- Cable-way penetrations
- Butt welds at K joints
- Tube-to-tube connections

In the design phase of the rigs, computations are performed to estimate the fatigue life of the loaded members and critical areas. It is important to the persons in charge on the rig to reduce the theoretical fatigue life by actions, such as asymmetrical loading, exceeding maximum allowable deck load, or unauthorized welding. The proper working of the cathodic protection should be verified at regular intervals and any corrosion around critical members should be reported.

TABLE 43.1

Requirements for Ordinary-strength Hull Structural Steel

PROCESS OF MANUFACTURE	FOR ALL GRADES: OPEN-HEARTH, BASIC-OXYGEN OR ELECTRIC-FURNACE					
GRADES	A	B	C	CS	D	E
DEOXIDATION	Any method[1]	Semi-killed or killed	Fully killed, fine-grain practice[10]	Fully killed, fine-grain practice[10]	Semi-killed or killed	Fully killed, fine-grain practice[10]
CHEMICAL COMPOSITION (Ladle analysis)						
Carbon, %		0.21 max.	0.23 max.[4]	0.16 max.	0.21 max.[5]	0.18 max.[5]
Manganese, %		0.80–1.10[2,6]	0.60–0.90[3]	1.00–1.35	0.60–1.40[6]	0.70–1.50[6]
Phosphorus, %	0.05 max.	0.05 max.	0.05 max.	0.05 max.	0.05 max.	0.05 max.
Sulphur, %	0.05 max.	0.05 max.	0.05 max.	0.05 max.	0.05 max.	0.05 max.
Silicon, %			0.10–0.35	0.10–0.35	0.35 max.	0.10–0.35
HEAT TREATMENT			Normalized over 35.0 mm (1.375 in.) thick.[8]	Normalized		Normalized
TENSILE TEST						
Tensile strength	For All Grades: 41–50 kgs per sq. mm[7,9] or 58000–71000 lbs. per sq. in. or 26–32 tons per sq. in.					
Elongation, min	For All Grades: 21% in 200 mm (8 in.)[7] (see 43.3.4c and 43.3.4d) or 24% in 50 mm (2 in.) (for specimen in Fig. 43.2) or 22% in 5.65 $\sqrt{A}$ (*A* equals area of test specimen)					
IMPACT TEST CHARPY STANDARD V-NOTCH						
Temperature					0C (32F)	−10C (14F)
Energy, avg., min.						
Longitudinal Specimens					4.8 kgm (35 ft. lbs.)	6.2 kgm (45 ft. lbs.)
or						
Transverse Specimens					3.2 kgm (23 ft. lbs.)	4.1 kgm (30 ft. lbs.)
No. of specimens					3 from each 40 tons	3 from each plate
STAMPING	AB/A	AB/B	AB/C (as rolled) AB/CN (normalized)	AB/CS	AB/D	AB/E

Notes

1 Unless otherwise specially approved, Grade A steel over 12.5 mm (0.50 in.) in thickness is to be semi-killed or killed with a minimum manganese/carbon ratio of 2.5. Grade A shapes and bars are not subject to the manganese/carbon ratio of 2.5.

2 Upper limit of manganese may be exceeded, provided carbon content plus ⅙ manganese content does not exceed 0.40%.

3 When the silicon content is 0.10% or more (killed steel) the minimum manganese content may be 0.60%.

4 For normalized Grade C plates, the maximum carbon content may be 0.24%.

5 Carbon content plus ⅙ manganese content shall not exceed 0.40%.

6 Where the use of cold-flanging quality has been specially approved (3.1), the manganese content may be reduced to 0.60–0.90%.

7 The tensile strength of cold-flanging steel shall be 39–46 kg/mm² (55000–65000 psi) and the elongation 23% min. in 200 mm (8 in.).

8 Grade C plates over 35.0 mm (1.375 in.) in thickness are to be ordered and produced in the normalized condition when intended for important structural parts.

9 A tensile strength range of 41–56 kg/mm² (58,000–80,000 psi) may be applied to Grade A shapes provided the carbon content does not exceed 0.26% by ladle analysis.

10 See 43.3.2d.

SECTION 43 Materials for Hull Construction and Equipment

Figure 8.4 ABS example of steel quality for ordinary hull structures

TABLE 43.2
Requirements for Higher Strength Hull Structural Steel

GRADES[1]—AH32, DH32, EH32

PROCESS OF MANUFACTURE Open Hearth, Basic Oxygen or Electric Furnace

CHEMICAL COMPOSITION (Ladle Analysis)

Carbon, % 0.18 max.	Manganese, %[2] 0.90–1.60	Phosphorus, % 0.04 max.	Sulfur, % 0.04 max.	Silicon, %[3] 0.10–0.40
Nickel, % 0.40 max.	Chromium, % 0.25 max.	Molybdenum, % 0.08 max.	Copper, % 0.35 max.	
Aluminum, %[4] (acid-soluble) 0.060 max.	Aluminum, %[4] (total) 0.065 max.			

TENSILE TEST

Tensile Strength	48–60 kg/mm^2; 68,000–85,000 psi; 30.5–38 tons per sq. in.
Yield Point, min.	32 kg/mm^2; 45,500 psi; 20.5 tons per sq. in.
Elongation, min.	19% in 200 mm (8 in.) or 22% in 50 mm (2 in.) (for specimen in Fig. 43.2) or 20% in 5.65 $\sqrt{A}$ (*A* equals area of test specimen)

TABLE 43.2 *(continued)*

HEAT TREATMENT See Table 43.4

	AH32	DH32	EH32
DEOXIDATION	Semi-killed or Killed[3]	Killed, Fine Grain Practice[5]	Killed, Fine Grain Practice[5]
IMPACT TEST STANDARD CHARPY V-NOTCH			
Temperature	None Required	−20C (−4F)	−40C (−40F)
Energy, min. avg.			
Longitudinal Specimens		3.5 kgm(25ft lbs)[6]	3.5 kgm(25ft lbs)
or			
Transverse Specimens		2.4 kgm(17ft lbs)[6]	2.4 kgm(17ft lbs)
No. of specimens		3 from each 40 tons	3 from each plate
STAMPING	AB/AH32	AB/DH32 or AB/DHN32[7]	AB/EH32

Notes

1. The numbers following the Grade designation indicate the yield point to which the steel is ordered and produced in kg/mm^2. A yield point of 32 kg/mm^2 is equivalent to 45,500 psi and a yield point of 36 kg/mm^2 is equivalent to 51,000 psi.
2. Grade AH 12.5 mm (0.50 in.) and under in thickness may have a minimum manganese content of 0.70%.
3. Grade AH to 12.5 mm (0.50 in.) inclusive may be semi-killed in which case the 0.10% minimum Silicon does not apply. Unless otherwise specially approved, Grade AH over 12.5 mm (0.50 in.) is to be killed with 0.10 to 0.40 percent Silicon.
4. Aluminum may be either partially or totally replaced by Columbium (Niobium) (0.05% maximum) or Vanadium (0.10% maximum) in which case the heat treating requirements of Table 43.4 for Columbium (Niobium)-treated or Vanadium-treated steel apply.
5. Grades DH and EH are to contain at least one of the grain refining elements in sufficient amount to meet the fine grain practice requirement. (See Paragraph 43.5.2(d)).
6. Impact tests are not required for normalized Grade DH.
7. The marking AB/DHN is to be used to denote Grade DH plates which have either been normalized or control rolled in accordance with an approved procedure.

SECTION 43 Materials for Hull Construction and Equipment

Figure 8.5 ABS example of steel quality for higher strength structures

8.11 LOAD CURVES ON DRILLING VESSELS

In addition to the usual computer stability calculation on drilling vessels, the program may automatically show a load curve with the corresponding shear and bending forces for sagging and hogging. The load curve shows the difference between the weight ordinate and the buoyancy ordinate. The computer calculates the shear force and bending moment curves corresponding to the load condition for sagging and hogging. The result of the shear force and bending moment curves should be within the allowable curves. Figure 8.6 is an example from a load condition with the shear force and bending moment curves, and Figure 8.7 is a table with the allowable still-water longitudinal forces and bending moments.

Table 3-4 – GOM Well Test, Operating Condition (Continued)

Loading Program 23.07.99 **Page 3**
LOADING REPORT **CONDITION: GOM TEST 3**
A/S CONSULTAS

LONGITUNAL STRENGTH:

	SHEAR FORCES			BENDING MOMENTS (-Hogg / + Sagg)		
Frame	tonnes	%Sea	%Harb.	tonnes.m	%Sea	%Harb.
99	2702	82	82	-13697	27	27
69	-667	18	18	-43872	36	36
65	-1901	31	31	-17335	9	9
61	-2320	37	37	23767	11	11
56	-286	5	5	56012	25	25
50	-1258	20	20	73893	33	33
44	864	14	14	87116	39	39
39	465	8	8	71579	39	39
33	1666	46	46	11637	12	12

Frame: 99 Max % Shear Force: 2702 tonnes 82 % Sea 82 % Harbour
Frame: 83 Max % Bending Moment: -39238 tonnes 44 % Sea 44 % Harbour

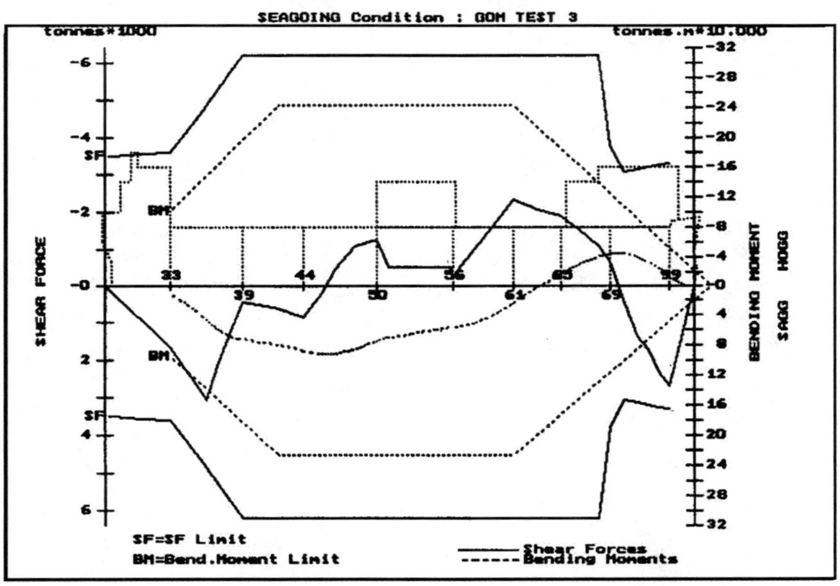

Figure 8.6 Example of shear force and bending moment curves

Table 2-5 – Allowable Still-Water Longitudinal Shear Forces and Bending Moments

Frame Nos.	Dist AP (m)	Shear Force (metric tons)	Bending Moments Hogging (metric ton-m)	Bending Moments Sagging (metric ton-m)
-7	-5.60	0	0	0
4	3.20	3,500	50,758	47,244
33	26.40	3,600	108,768	101,237
36	41.25	4,850	123,270	114,735
39	56.10	6,200	138,226	128,655
42	70.95	6,200	242,915	226,096
61	166.25	6,200	242,915	226,096
68	200.90	6,200	138,226	128,655
69	205.85	3,763	123,270	114,735
75	210.65	3,044	108,768	101,237
99	229.85	3,300	50,758	47,244
120	246.65	0	0	0

Figure 8.7 Maximum shear forces and bending moments

8.12 PRELOADING ON A JACKUP

Once the jackup is on position, the legs are lowered to the seabed and from thereon supported by the spud cans in contact with the soil. In addition to the weight of the rig on the legs, the rig is exposed to the horizontal forces by wind, wave, and current. These forces tend to overturn the rig with an additional vertical load on the leeward leg. If, for any reason, the soil does not support this additional load in case of a storm, additional penetration of the leeward leg will result in tilting of the rig (fig. 8.8). The result will be excessive deformation or even fracture of the leg.

To prevent such a catastrophic structure failure, the rig will be preloaded by filling a set of preload tanks with seawater. The purpose of preloading is to simulate the maximum reaction that the rig would experience in case of a storm. This will ensure that the soil underneath the spud cans will support the rig with all the additional loads caused by the environment.

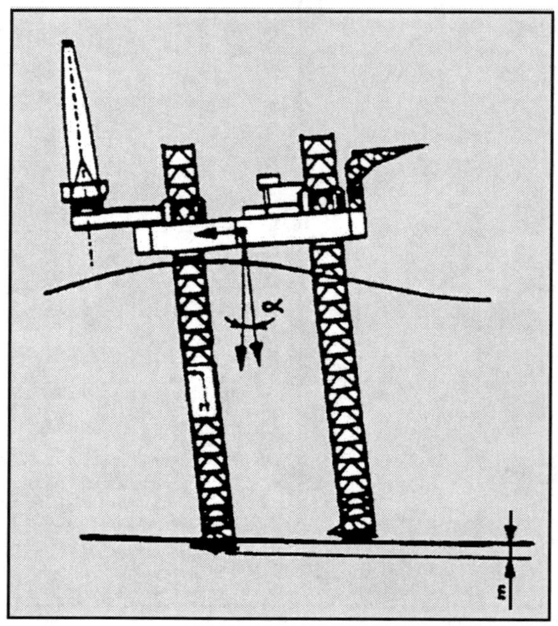

Figure 8.8 Jackup tilting due to penetration of one leg

8.13 THE STRENGTH OF THE JACKUP STRUCTURE

The jackup in elevated position resists the environmental forces as a flexible structure with the support at the spud cans or mat. The structure combination from spud cans, legs, jackhouse and the hull divided by bulkheads forms a space frame. The entire frame is subjected to loads. The stresses within the various members should be maintained within the allowable limits.

8.14 THE FORCES ON THE JACKUP

The forces computed to check the strength of the jackup structure are the

1. storm forces from three directions.
2. vertical forces on the legs.
3. horizontal forces or shear forces on the legs.
4. bending moment on the legs.

8.15 STRUCTURAL ELEMENTS OF THE JACKUP

For a jackup there are basically categories of structural members.

1. *Special areas.* These are the primary structural elements:
   - ❖ The connections of the legs to the spud can or mat.
   - ❖ The jackhouse connection to the hull.
   - ❖ The hold-down claw of the cantilever.
2. *Primary structural elements.* These are elements essential to the overall integrity of the rig.
   - ❖ Leg tubulars.
   - ❖ Hull girders.
   - ❖ Jackhouse members.
   - ❖ Crane pedestals.
   - ❖ Helideck connection to hull.
3. *Secondary structural elements.* These elements are of minor importance. Failure will not affect the integrity of the rig. The pictures in Figure 8.9 show some examples of the structural classification

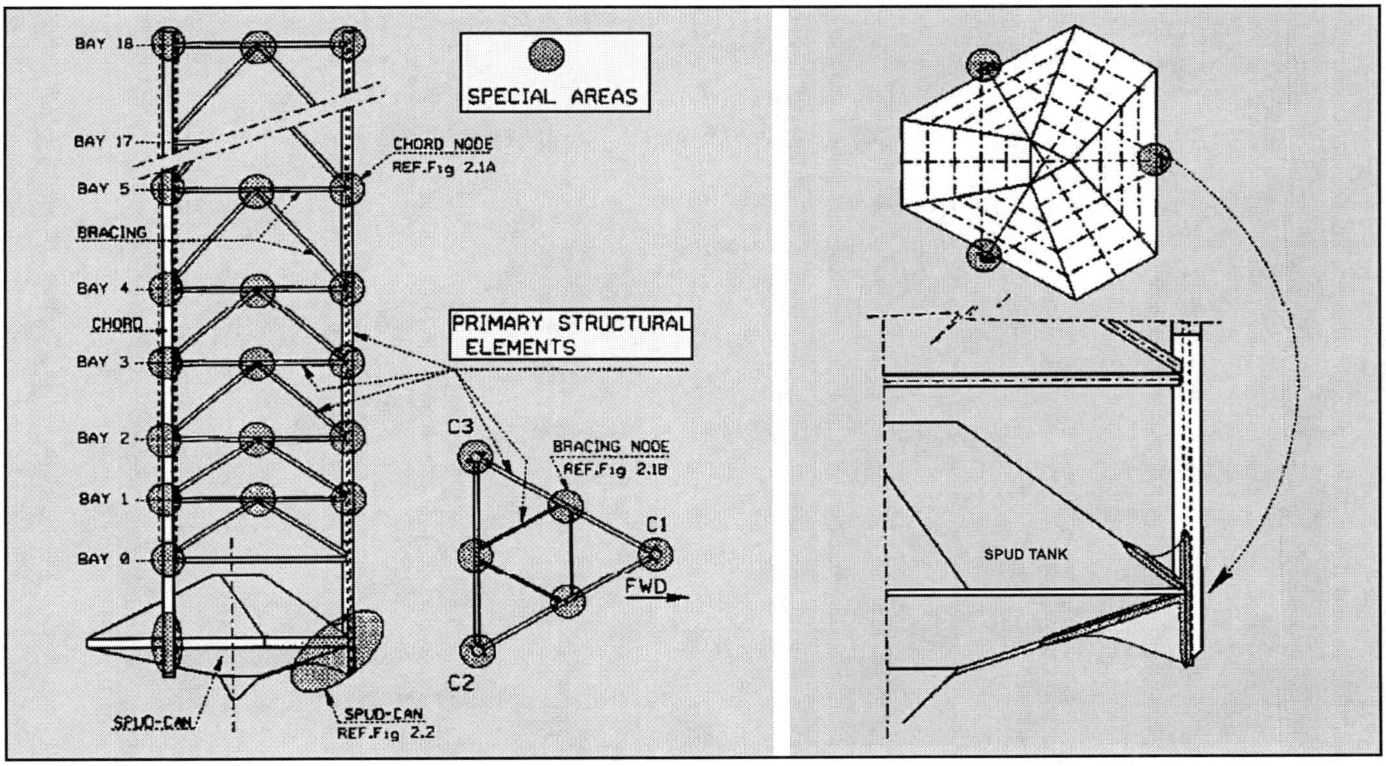

Figure 8.9 Display of leg and spud can structural classification

EXERCISE
Chapter 8

1. Explain what prying loads are.
2. Name at least five areas on the rig you work on with high stresses and high fatigue tendency.
3. Why must a jackup preload before starting operations?
4. Name the two main categories of forces (loads) the rig is subjected to.
5. Explain the principle of fatigue.
6. Explain the difference between stress and strain.
7. What is sagging and hogging?
8. What is the most critical structural element of a jackup?
9. What does the load curve of a drilling vessel represent?
10. Explain what a utilization factor of .75 represents.
11. Give the definition of the yield point.
12. At what point will a steel structure deform permanently?
13. Explain what half-gravity loads are.
14. Under what condition is it allowed to have a limited amount of asymmetric loading applied to the rig?
15. Name at least one force that will cause twisting.
16. What is endurance limit?

ANSWER KEY
Advanced Stability

CHAPTER 1

Exercise 1, page 1.8

1. Center of gravity is the point through which the total weight of body is considered to be acting downward.

2. Mass is the physical matter in the body, weight is the vertical force experienced by a mass as a result of gravity.

3. a. *Resultant force value* = 240 – 110 = 130 and 180 – 95 = 85

$130^2 + 85^2 = \sqrt{16{,}900 + 7{,}225}$
$= \sqrt{24{,}125}$
$= \mathbf{155.32\ kg}$

Resultant force angle – sin 85/155 = .548 = **33.25°**

b. *Resultant force value* = $6.4^2 + 3.8^2 = \sqrt{40.96 + 14.40}$
$= \sqrt{55.36}$
$= \mathbf{7.44\ st}$

Resultant force angle – sin 3.8 ÷ 7.44 = .51 = **30.66°**

4. *Unknown distance*

$$= \frac{[(5 \times 0) + (13.3 \times 3.5) + (17.5 \times 10.8)]}{(5 + 13.3 + 17.5)}$$
$$= \frac{235.55}{35.8}$$
$$= \mathbf{6.57\ ft}$$

5. *Unknown weight*

Total clockwise moments = Total anticlockwise moments

$(x \times 6) + (8.5 \times 2.5) = (8.3 \times 2.5) + (7.2 \times 5)$
$(6x) + (21.25) = (20.75) + (36)$
$(6x) = (56.75) - (21.25)$
$x = 35.5 \div 6$
$x = \mathbf{5.91\ kg}$

6. *Unknown distance*

$$= \frac{[(69.8 \times 275) + (39.3 \times 650) + (0 \times 850)]}{(275 + 650 + 850)}$$
$$= \frac{44{,}740}{1{,}775}$$
$$= \mathbf{25.2\ m}$$

7. *Distance LCG*

$$= \frac{[(4.6 \times 41) + (7.7 \times 16) + (12.1 \times 62) + (6.3 \times 203)]}{(41 + 16 + 62 + 203)}$$
$$= \frac{2{,}340.9}{322}$$
$$= \mathbf{7.26\ m}$$

Exercise 2, page 1.28

1. *Density* = weight ÷ volume
= 1,900 ÷ (9.2 × 1.6 × .7)
= **184.39 kg/m³**

2. *Weight* = volume × density
= 55 × 64
= **3,520 lb**

3. *Weight* = volume × density
= 5,000 × 1,025
= 5,125,000 kg
= 5,125,000 ÷ 1,000
= **5,125 mt**

4. *Volume* = weight ÷ density
= (12,400 × 2,000) ÷ 62.5
= **396,800 ft**

5. *Relative density* = density of substance ÷ density of fresh water
= 300/1,000
= **.3 kg/m**

6. *Volume* = weight ÷ density
= (22,500 × 2204.6) ÷ 64
= **775,054 ft**

7. *Weight* = volume × density
= 598,000 × 64
= 38,272,000 lb
= 38,272,000 ÷ 2,000
= **19,136 st**

8. *Volume* = weight ÷ density
= (2,750 × 2,000) ÷ 64
= **85,937.5 ft³**
Draft = volume ÷ (length × width)
= 85,937.5 ÷ (320 × 37)
= **7.25 ft draft in seawater**
New draft = old density ÷ new density) × old draft
= (64 ÷ 62.5) × 7.25
= **7.42 ft**
Change in draft = 7.42 – 7.25
= **.17 ft**

9. *Volume* = weight ÷ density
= (555 × 1,000) ÷ 1,025
= **541.46 m**
Draft = volume ÷ (length × width)
= 541.46 ÷ (45 × 9.1)
= **1.32 m**

10. *COD* = (W ÷ TPC_{sw}) × [(old density – new density) ÷ new density]
= (33,000 ÷ 7.04) × [(1,025 – 1,015) ÷ 1,015]
= 46 cm
New draft = 18 + .46
= **18.46 m**

11. *New displacement* = old displacement × new density ÷ old density
= 13,500 × 1.002 ÷ 1.025
= 13,197 mt
Cargo to be removed = 13,500 – 13,197
= **303 mt**

12. *Water plane area* = length × width
= 250 × 35
= 8,750 m^2
TPC = (1.025 × 8,750) ÷ 100
= **89.68 t/cm**

13. Feet: **46' – 46' 9" – 47'6"**
Meters: **17.80 m – 17.95 m – 18.10 m**

14. a. *Volume* = weight ÷ density
= (13,000 × 2,000) ÷ 64
= 406,250 ft^3
Draft = volume ÷ (length × width)
= 406,250 ÷ (445 × 72)
= 12.67 ft (12'8") draft in seawater
Freeboard = 27 – 12.67 = **14.32 ft (14'3.8")**

b. *New displacement* = 13,000 + 5,500 = 18,500 st
Volume = weight ÷ density
= (18,500 × 2,000) ÷ 64
= 578,125 ft^3
Draft = volume ÷ (length × width)
= 578,125 ÷ (445 × 72)
= 18.05 ft draft in seawater
Freeboard = 27 – 18.05 = **8.96 ft (8'11.5")**

c. *Reserve buoyancy* = length × width × freeboard
= 445 × 72 × 8.96
= **287,078 ft**

d. TPI = WPA ÷ 375 $\quad 375 = \dfrac{2{,}000 \times 12}{64}$
= (445 × 72) ÷ 375
= **85.44 st/in.**

e. *New draft* = old draft × old density ÷ new density
= 18.04 × 64 ÷ 63
= **18.32 ft (18'3.8")**

f. *Displacement* = length × width × draft
= 445 × 72 × 18.32
= **586,972 ft³**

Exercise 3, page 1.33

1. *Initial mean draft* = (14 + 10) ÷ 2 = 12
Final mean draft = (13 + 9) ÷ 2 = 11
COD = 12 – 11 = **1 ft (12")**

Initial trim = 14 – 10 = 4 ft down on aft
Final trim = 13 – 9 = 4 ft down on aft
COT = 4 – 4 = **0**

2. *New draft aft* = 12.30 + (.42 ÷ 2) = **12.51 m**

New draft fwd = 12.30 – (.42 ÷ 2) = **12.09 m**

3. a. *The COT*
Initial trim = 22' 6" – 19' 6" = 3 ft down on aft
Final trim = 21' 6" – 19' 4" = 2' 2" down on aft
COT = 3 – 2'2" = **10" (.83 ft) by the bow**

b. *The change of mean draft*
Initial mean draft = (22' 6" + 19' 6") ÷ 2 = 21'
Final mean draft = (21' 6" + 19' 4") ÷ 2 = 20'5"
COD = 21' – 20'5" = **7" (.58 ft)**

4. *Initial draft fwd and aft is 6.35 m*
COT fwd = trim × length fwd ÷ total length
= .70 × 84.70 ÷ 160 (84.70 = 160 ÷ 2 + 4.70)
= .37 m

COT aft = trim × length aft ÷ total length
= .70 × 75.30 ÷ 160 (75.30 = 160 ÷ 2 – 4.70)
= .33 m

Final draft fwd = 6.35 – .37 = **5.98 m**

Final draft aft = 6.35 + .33 = **6.68 m**

True mean draft = final draft fwd + (trim × length fwd ÷ total length)
= 5.98 + (.70 × 84.70 ÷ 160)
= **6.35 m**

5. a. The initial trim
Initial trim = 16'6" – 15'6" = 1 ft (12") down on bow

b. The initial mean draft amidships
Initial mean draft = (16' 6" + 15' 6") ÷ 2 = 16'

c. The initial true mean draft
True mean draft = draft fwd ± (trim × length fwd ÷ total length)
= 16.5 – (1 × 230 ÷ 420)
= **15.96 ft (15'11.5")**

d. The final trim
Initial trim = 12" on bow
COT = 20" calculated on aft
Final trim = 20" – 12" = **8" (.66 ft) down on aft**

e. The final draft forward and aft
COT fwd = trim × length fwd ÷ total length
= 1.66 × 230 ÷ 420
= .90 ft (11")
COT aft = trim × length aft ÷ total length
= 1.66 × 190 ÷ 420
= **.75 ft (9")**

Final draft fwd = 16'6" – 11" = 15'7" (15.58 ft)
Final draft aft = 15'6" + 9" = **16'3" (16.25 ft)**

f. The COT
Initial trim = 16' 6" – 15' 6" = 1 ft down on bow
Final trim = 16'3" – 15'7" = 8" down on aft
COT = 1' + 8" = **20" (1.66 ft)**

g. The final mean draft amidships
Final mean draft = (15'7" + 16'3") ÷ 2 = **15'11" (15.91 ft)**

h. The final true mean draft
True mean draft = final draft fwd ± (Trim × length fwd ÷ total length)
= 15.58 + (.66 × 230 ÷ 420)
= **15.94 ft (15'11.2")**

i. The change of displacement
Zero

6. a. The initial trim
Initial trim = 17.00 m – 16.75 m = **.25 m down on aft**

b. The initial mean draft
Initial mean draft = (17.00 m + 16.75 m) ÷ 2 = **16.875 m**

c. The initial true mean draft
True mean draft = draft fwd ± (trim × length fwd ÷ total length)
= 16.75 + (.25 × 50.7 ÷ 105)
= **16.87 m**

d. The final draft forward and aft
COT fwd = trim × length fwd ÷ total length
= .40 × 50.7 ÷ 105
= .19 m
COT aft = trim × length aft ÷ total length
= .40 × 54.3 ÷ 105
= .21 m
Final draft fwd = 16.75 + .19 = **16.94 m**
Final draft aft = 17.00 – .21 = **16.79 m**

e. The COT
Initial trim = 17.00 m – 16.75 m = .25 m down on aft
Final trim = 16.94 m – 16.79 m = .15 m down on bow
COT = .25 + .14 = **.40 m**

f. The final trim
Final trim = 16.94 m – 16.79 m = **.15 m down on bow**

g. The final mean draft
Final mean draft = (16.94 m + 16.80 m) ÷ 2 = **16.87 m**

h. The final true mean draft
True mean draft = draft fwd ± (trim × length fwd ÷ total length)
= 16.94 – (.15 × 50.7 ÷ 105)
= **16.87 m**

7. a. 1) The initial mean draft
Initial mean draft = (23'08" + 23'04") ÷ 2 = 23'6" (23.5 ft)

2) The initial true mean draft
Initial trim = 23'08" – 23'04" = 04" down on bow
True mean draft = draft fwd ± (trim × length fwd ÷ total length)
= 23.66 – (.33 × 300 ÷ 500)
= **23.42 ft (23'5")**

b. 1) The change of trim
COT = 208 × 328 ÷ 6,200 = **11" (.91 ft)**

2) The final forward and aft drafts
COT fwd = trim × length fwd ÷ total length
= .91 × 300 ÷ 500
= **.54 ft (6.5")**
COT aft = trim × length aft ÷ total length
= .91 × 200 ÷ 500
= **.36 ft (4.32")**

Final draft fwd = 23'08" – 6.5" = 23'1" (23.08 ft)
Final draft aft = 23'04" + 4.32" = 23'8" (23.66 ft)
Trim = 23'1" – 23'8" = 7" (.58 ft) down on aft

3) The final mean draft
Final mean draft = (23'1" + 23'8") ÷ 2 = **23'4.5" (23.37 ft)**

4) The final true mean draft
Final true mean draft = draft fwd ± (trim × length fwd ÷ total length)
= 23.125 + (.58 × 300 ÷ 500)
= **23.47 ft (23'5.6")**

8. a. The initial trim
Initial trim = 8.22 m – 8.60 m = **.38 down on aft**

b. The initial mean draft
Initial mean draft = (8.60 + 8.22) ÷ 2 = **8.41 m**

c. The initial true mean draft
True mean draft = draft fwd ± (trim × length fwd ÷ total length)
= 8.22 + (.38 × 58.7 ÷ 160)
= **8.42 m**

d. The final draft forward and aft
COT due to adding load = 160 × 15.3 ÷ 115 MCT = **21.2 cm (.21 m)**
COT fwd = trim × length fwd ÷ total length
= .21 × 58.7 ÷ 114
= .11 mt

COT aft = trim × length aft ÷ total length
= .21 × 55.3 ÷ 114
= .10 mt

Draft increase due to adding load = 160 mt ÷ 17 TPC = 9.41 cm
= .09 mt

Final draft fwd = 8.22 + .09 – .11 m = **8.20 m**
Final draft aft = 8.60 + .09 + .10 m = **8.79 m**
Trim = **.59 m down on aft**

e. The COT
Initial trim = .38 down on aft
Final Trim = .59 m down on aft
Change = .59 – .38 = **.21 m more on aft**

f. The final trim
Final trim = **.59 m down on aft**

g. The final mean draft
Final mean draft = (8.20 m + 8.79 m) ÷ 2 = **8.49 m**

h. The final true mean draft
Final true mean draft = draft fwd ± (trim × length fwd ÷ total length)
= 8.20 + (.59 × 58.7 ÷ 114)
= **8.50 m**

9. *Moments:*
+161 × (295 – 220) = +12,075
+143 × (185 – 220) = -5,005
-180 × (265 – 220) = -8,100
-124 × (85-220) = +16,740
Increase in displacement = 0 = +15,710 *total moments*

COT = total moments ÷ MT1"
= 15,710 ÷ 1,570 = 10" (.83 ft)

COT fwd = trim × length fwd ÷ total length
= .83 × 180 ÷ 400
= .37 ft (4.5")
COT aft = trim × length aft ÷ total length
= .83 × 220 ÷ 400
= .45 ft (5.5")

Final draft fwd = 16'08" + 4.5" = 17'0.5" (17.04 ft)
Final draft aft = 18'05" – 5.5" = 17'11.5" (17.95 ft)

10. *Moments:* +176 × (226 – 124) = +17,952

Total moments for loading = +17,952
Increase in displacement = 176
Draft increase = 176 ÷ 44 = 4"

COT = total moments ÷ MT1"
= 17,952 ÷ 2,400 = 7.48"(.62 ft)

COT fwd = trim × length fwd ÷ total length
= .62 × 256 ÷ 482
= .33 ft (4")

COT aft = trim × length aft ÷ total length
= .62 × 226 ÷ 482
= .29 ft (3.5")

Final draft fwd = 25'2" + 4" – 4" = **25'2" (25.16 ft)**
Final draft aft = 25'6"+ 4" + 3.5" = **26'1.5" (26.125 ft)**

Exercise 4, page 1.44

1. Tan 0 = opposite side ÷ adjacent side
Tan 0 = .416 ÷ 6 .416 = 5 ÷ 12
Tan 0 = .0693 = **3.96°**

2. $$GM = \frac{w \times d}{W} \times \frac{\text{length of pendulum}}{\text{deflection of pendulum}}$$

$$GM = \frac{12 \times 49}{6{,}360} \times \frac{480}{14} = .092 \times 34.28 \quad (480 = 40 \times 12)$$

GM = 3.15 ft
KM = KG + GM so KG = KM – GM
KG = 26.88 – 3.15 = **23.73 ft (23'8.7")**

3. $$GM = \frac{w \times d}{W} \times \frac{\text{length of pendulum}}{\text{deflection of pendulum}}$$

$$GM = \frac{32 \times 36}{34{,}120} \times \frac{7.9}{.089} = .0337 \times 88.76$$

GM = 2.99 m
KM = KG + GM so KG = KM – GM
KG = 27 – 2.99 = **24.01 m**

Lightship KG

Weight	KG	Moments
34,120	24.01	819,221
-750	2.60	-1,950
-950	26.8	-25,460
-120	24.9	-2,988
32,300		788,823

Lightship displacement = 32,300 mt
Lightship KG = moments / displacement
= 788,823 / 32,300
= 24.42 m
Lightship GM = KM – KG
= 27.80 – 24.42
= **3.38 m**

4. $$GM = \frac{w \times d}{W} \times \frac{\text{length of pendulum}}{\text{deflection of pendulum}}$$

$$W = \frac{w \times d}{GM} \times \frac{\text{length of pendulum}}{\text{deflection of pendulum}}$$

$$W = \frac{12 \times 10.8}{.75} \times \frac{8.20}{.2} = 172.8 \times 41 = \mathbf{7{,}048.8\ mt}$$

5. a. *For the tank without division*

$$GG_v = \frac{I}{V} \times \frac{\text{density of fluid in tank}}{\text{density of water}}$$

$$I = \frac{L \times B^3}{12} \qquad V = W / d$$

$$I = \frac{60 \times 54^3}{12} = 787{,}320$$

$$V = \frac{8{,}190 \times 2{,}000}{64} = 255{,}937.5$$

$$GG_v = \frac{I}{V} \times \frac{\text{density of fluid in tank}}{\text{density of water}}$$

$$GG_v = \frac{787{,}320}{255{,}937} \times \frac{64}{64} = \mathbf{3.07\ ft}$$

b. For the tank with a longitudinal centerline bulkhead

$$GG_v = \frac{I}{V} \times \frac{\text{density of fluid in tank}}{\text{density of water}}$$

$$I = \frac{L \times B^3}{12} \qquad V = W / d$$

$$I = \frac{60 \times 27^3}{12} = 98{,}415$$

$$V = \frac{8{,}190 \times 2{,}000}{64} = 255{,}937.5$$

$$GG_v = \frac{I}{V} \times \frac{\text{density of fluid in tank}}{\text{density of water}}$$

$$GG_v = \frac{98{,}415}{255{,}937} \times \frac{64}{64} = .384 \text{ ft for two tanks}$$

$$= 2 \times .384 = \mathbf{.770\ ft}$$

6. Total free surface moments

Tank	Free surface moments	
PB1A = 9 ft	1,286	
SB1B = 2 ft	645	Interpolation (833 – 357/2 + 357) = 645
DW1 = 19 ft	1,840	Interpolation (2,015 – 1,664/2 + 1,664) = 1,840
DW2 = 20.5 ft	832	Interpolation (1,664 – 0/2 + 0) = 832
SB10B = 1 ft	457	
PB10B = 4 ft	894	Interpolation (1,017 – 833/3 + 833) = 894
Total free surface moments	5,954	

$$KG \text{ wet} = \frac{\text{free surface moments}}{\text{displacement}} + KG$$

$$KG \text{ wet} = \frac{5{,}954}{25{,}700} + 61.4 = 61.63$$

$$\begin{aligned} GM \text{ wet} &= KM - KG \text{ wet} \\ &= 67 - 61.63 \\ &= \mathbf{5.37\ ft\ (5'4.4")} \end{aligned}$$

7. *Angle of heel without free surface*

$$GG' = \frac{w \times d}{\text{displacement}}$$

$$GG' = \frac{420 \times 15}{18{,}500} = .34 \text{ m}$$

Tan 0 = opposite side / adjacent side
Tan 0 = .34 / 3.4
Tan 0 = .10 = **5.71°**

Angle of heel with free surface

$$GG_v = \frac{I}{V} \times \frac{\text{density of fluid in tank}}{\text{density of water}}$$

$$I = \frac{L \times B^3}{12} \qquad V = W / d$$

$$I = \frac{12 \times 14^3}{12} = 2{,}744$$

$$V = \frac{18{,}500}{1.025} = 18{,}048.78$$

$$GG_v = \frac{I}{V} \times \frac{\text{density of fluid in tank}}{\text{density of water}}$$

$$GG_v = \frac{2{,}744}{18{,}048} \times \frac{1.025}{1.025} = \mathbf{.15\ m\ for\ two\ tanks}$$

$$= 2 \times .15 = \mathbf{.30\ m}$$

GM Wet = 3.4 – .30 = 3.10
Tan 0 = opposite side / adjacent side
Tan 0 = .34 / 3.10
Tan 0 = .109 = **6.22°**

Exercise 5, page 1.48

1. BM = KM – KB
KB = ½ draft
KB = .5 × 19 = 9.5 ft
BM = 22.5 – 9.5 = 13 ft

2. *Shift of G transverse due to weight addition*

$$GG' = \frac{w \times d}{\text{displacement} + w}$$

$$GG' = \frac{24 \times 42}{18{,}440 + 24} = .054 \text{ m}$$

Shift of G vertical due to weight addition

$$GG' = \frac{w \times d}{\text{displacement} + w}$$

$$GG' = \frac{24 \times 44.2}{18{,}440 + 24} = .057 \text{ m}$$

$$\text{New KG} = 15.8 + .057 = 15.857 \text{ m}$$

$$\text{GM} = \text{KM} - \text{KG}$$
$$= 21 - 15.857 = 5.14 \text{ m}$$

$$\text{Tan } 0 = \text{opposite side / adjacent side}$$
$$\text{Tan } 0 = .054 / 5.14$$
$$\text{Tan } 0 = .10 = \mathbf{.57°}$$

3. $\text{KM} = \text{KB} + \text{BM}$

$$\text{KB} = \tfrac{1}{2}\text{ draft}$$
$$= .5 \times 6.8 = 3.4 \text{ m}$$

$$\text{BM} = \frac{I}{V}$$

$$I = \frac{L \times B^3}{12} \qquad V = l \times W \times d$$

$$I = \frac{130 \times 20^3}{12} = 86{,}666$$

$$V = 130 \times 20 \times 6.8 = 17{,}680 \text{ m}^3$$

$$\text{BM} = \frac{I}{V} = \frac{86{,}666}{17{,}680} = 4.90 \text{ m}$$

$$\text{KM} = \text{KB} + \text{BM}$$
$$= 3.4 + 4.9$$
$$= \mathbf{8.3 \text{ m}}$$

4. *New KG*

Weight	KG	Moments
13,225	12.34	163,196
+350	14.5	+ 5,075
+1,250	5	+6,250
+30	25	+750
+950	15.5	+14,725
+150	12.1	+1,815
-560	13.25	-7,420
-330	6.01	-1,983.30
-520	16.90	-8,788
-150	11.12	-1,668
14,395		171,951.7

New KG = moments / weight
= 171,951 / 14,395 = **11.94 m**

5. $\text{KM} = \text{KB} + \text{BM}$

$$\text{KB} = \tfrac{1}{2}\text{ draft}$$
$$= .5 \times 3.5 = 1.75 \text{ m}$$

$$\text{BM} = \frac{I}{V}$$

$$I = \frac{L \times B^3}{12} \qquad V = l \times W \times d$$

$$I = \frac{75 \times 12^3}{12} = 10{,}800 \text{ m}^4$$

$$V = 75 \times 12 \times 3.5 = 3{,}150 \text{ m}^3$$

$$\text{BM} = \frac{I}{V} = \frac{10{,}800}{3{,}150} = 3.42 \text{ m}$$

$$\text{KM} = \text{KB} + \text{BM}$$
$$= 1.75 + 3.42$$
$$= \mathbf{5.17 \text{ m}}$$

$$\text{KM} = \text{KB} + \text{BM}$$
$$\text{KB} = \tfrac{1}{2}\text{ draft}$$
$$\text{KB} = 0.5 \times 3.5 = 1.75$$

$$\text{BM} = \frac{I}{V}$$

$$\text{BM} = \frac{LB^3}{12V}$$

$$\text{BM} = \frac{75 \times 13.8 \times 13.8 \times 13.8}{12 \times 75 \times 13.8}$$

$$\text{KM} = \text{KB} + \text{BM}$$
$$= 1.75 + 4.53$$
$$= 6.73 \text{ m}$$

Increase of KM transverse = 6.28 – 5.17 = **1.11 m**

6. $\text{Tan } 4° = .069$

$$\text{Tan } 0 = \text{opposite side / adjacent side}$$
$$.069 = \text{opposite side} / 3.8 \; (3.8 = 16.4 - 12.6)$$
$$\textit{Opposite side} \text{ (TCG)} = .069 \times 3.8$$
$$= .262 \text{ ft}$$

Weight	KG	Vmoments	TCG	Tmoments
4,800	12.6	60,480	.262	1,257
+80	16.5	+1,320	-12.0	-960
+40	20.4	+816	+26	+1,040
-60	4.2	-252	-8	+480
-60	2.4	-144	+15	-900
4,800	**12.96**	**62,220**	**.19**	**917**

Tan 0 = opposite side / adjacent side
Tan 0 = .19 / 3.44 (3.44 = 16.4 – 12.96)
= .055
= **3.16°**

Exercise 6, page 1.49

1. **The geometrical center of the underwater shape of the vessel.**

2. **Rig Type 1 is symmetric port/starboard.**

3. The height of the metacenter is the vertical distance between the keel and the metacenter. Metacentric height is the distance between the center of gravity and the metacenter.

4. Metacenter is a point of intersection between a line drawn vertically through the center of buoyancy and gravity in a level condition and a line drawn vertically through the center of buoyancy when the vessel is inclined.

5. The larger the GM, more stable the vessel.

6. Because the resistance to rotate along the shorter transverse waterplane is less than the resistance to rotate along the longer longitudinal waterplane.

7. When the vessel is symmetrical along both the axes.

8. Two equal and opposite parallel forces that form a restoring force that upright a vessel when the overturning force is removed.

9. The displacement of the vessel and the righting arm GZ

10 b. Center of gravity

11. It is the tipping center of the vessel. The point about which the rig heels or trims.

12. The water plane area

13. GM = -1 (unstable)

14. *Stiff*—quick and unpleasant movements (large stability couple)

    Tender—lazy and slow (small stability couple)

15. G will stay in its same position. B will move to the low side of the inclined vessel.

16. a. *Stable equilibrium.* Tendency of a vessel to return to upright. GM is positive.
    b. *Neutral equilibrium.* Condition where a vessel stays in the inclined position after an external force is removed. GM is zero.
    c. *Unstable equilibrium.* Condition where a vessel continues to incline afer an external force is removed. GM is negative.

CHAPTER 2

Exercise 1, page 2.19

1. a. roll and surge
   b. heave and yaw
   c. pitch and sway

2. a. yaw, pitch, roll
   b. surge, sway, heave

3. Wave crest velocity = length ÷ time
   = 400 ÷ 11
   = **36.36 ft/sec**

4. Wave velocity in kts = 36.36 × 360 (360 sec in an hour) ÷ 6,076
   = **21.54 kt**

5. Wave frequency = 1 ÷ time
   = 1 ÷ 11
   = .09 waves/sec
   = .09 × 60 = **5.4 waves/min**

6. Wave velocity = H ÷ 2 × angular frequency
   = 42 ÷ 2 × .495
   = **10.39 ft/sec**

7. Wave velocity in kts = 10.39 × 360 ÷ 6,076
   = **6.15 kt**

8. angular frequency (ω) = 2π ÷ T

   where

   T = period in seconds
   π = 3.14
   ω = .495 rad/sec (from question 6)

   therefore

   T = 2π ÷ ω
   = 2 × 3.14 ÷ .495
   = **12.68 seconds**

9. about 23 feet

Exercise 2, page 2.20

1. a. 0.3 (from chart) × 25 ft = 7.5 ft heave
   b. 0.2 (from chart) × 8 ft = 1.6 degrees

2. b. Increase the displacement

$$T = 2\pi\sqrt{\frac{W + W_a}{g \times \delta \times A}}$$

where

$W + W_a$ = displacement and added mass
g = acceleration of gravity
δ = density of seawater
A = waterplane area

Increasing the diameter of the column will increase "A" in the formula, resulting in a decrease in heave period.

Increasing the displacement will increase "W" above in the formula which will result in an increase in heave period.

Decreasing the width of the pontoons to reduce the added mass will decrease W_a in the formula above resulting in a decrease in heave period.

3. Wind force F = .00338 × C_h × C_s × V^2 × A

where

C_h = height coefficient from table 1 = 1.10 for 55 ft
C_s = shape coefficient from table 2 = 1.25
V = 75 kts
A = area = (8 × 125) × .6 open truss work = 600 ft

therefore

F = .00338 × 1.10 × 1.25 × 75^2 × 600
= **15,685.3 ft/lbs**

4. If the wave period and the natural period of the rig are too close, the wave action will cause excessive synchronous motions.

What actions can be taken?

Decrease W or GM because the formula to calculate natural pitch and roll is:

$$T = 2\pi\sqrt{\frac{I}{W \times GM}}$$

where

I = total inertia of the rig including the added mass *(The added mass is the amount of water that moves with the rig. This amount can be considerable and depends on the shape and direction.)*
W = the displacement of the rig
GM = the metacentric height

CHAPTER 3

Exercise 1, page 3.16

1. *Old draft* = 19.10
Draft change due to weight loaded = MT ÷ TPC
= 410 ÷ 34.5
= 11.88 cm (.12 m)
New draft = old draft + draft change
= 19.10 + .12 cm
= **19.22 cm**

2. *Moments* = Wt × distance
= 410 × 245 = 10,0450
COT = moments ÷ MT1"
= 10,0450 ÷ 7190 = 14"
Total Trim = **14"**
Therefore each side fwd and aft = 14 ÷ 2 = **7"**

New draft fwd = old draft + trim (plus because weight is shifted forward causing draft to increase)
= 25'4" + 7" = **25'11"**
New draft aft = old draft – trim (minus because weight is shifted forward causing draft to decrease)
= 27'2" – 7" = **26'7"**

3. Wt × distance = moments
Loaded + 60 × +(13 + 2.2) = 912
(Note 2.2 is added cause LCF is 2.2 aft of amidship)
Loaded + 210 × -(37.8) = -7938 (116 ÷ 2 – 2.2 – 18 = 37.8)
Discharged – 55 × -(18 -2.2) = +869
(Note 2.2 is subtracted cause LCF is 2.2 aft of amidship)

Increase in displacement = 215
Moments = -6,157 (aft)
COT = moments ÷ MCT 1 cm
= 6,157 ÷ 112 = 54.9 cm (.55 m)
Increase in draft = 215 ÷ 18 = **11.94 cm (.12m)**

COT fwd = trim × length fwd from LCF ÷ total length
= .55 × 60.2 ÷ 116
= **.28 cm** (minus)

COT aft = trim × length aft from LCF ÷ total length
= .55 × 55.8 ÷ 116
= **.27 cm** (plus)

New drafts	FWD	AFT	
Old draft	8.30	8.80	
Increase in draft	+ .12	+ .12	
COT	- .28	+ .27	*(+ cause moments were towards aft)*
	8.14 m	9.19 m	

4. Displacement at 20.9 m draft = 35,103 t
Displacement at 20.04 m draft = 34,287 t
Increase in displacement = 35,103 – 34,287
= **605 tonnes**

5. Displacement at 20.60 m draft = 34,893 t
Displacement at 20.04 m draft = 34,287 t
Tonnes that need to be discharged to change the draft = 34,893 – 34,287
= **606 tonnes**

6. Displacement = 34,850 t
Weight discharged = 834 t
New displacement = 34,850 – 834
= **34,016 tonnes**
New average draft at displacement 34,016 mt = **19.36 m**

7. a. Displacement = 34,365 tonnes
b. VCB = 6.99 m
c. LCB = 52.95 m
d. TCB = 0
e. WPA = 684 m^2
f. TPC = 7.04 t/cm
g. LCF = 52 m
h. TCF = 0
i. KM_L = 23.42 m
j. KM_T = 22.83 m
k. MCT = 75.5 t-m/cm
l. MCH = 114 t-m/cm

CHAPTER 3

Exercise 2, page 3.17

1. @ 70 ft draft KG = 60 and MH1° = 3,226
Moment = WT × distance
= 40 × (95.5 + 95.5)
= 7,640
Heel = moment ÷ MH1°
= 7,640 ÷ 3,226
= 2.3 degrees

2. @ 72 ft draft KG = 61.95 ft
New KG
Wt × VCG = V- moments
26,235 × 61.95 = 1,625,258
200 × 130 = 26,000
26,435 (new displacement) and draft = 73.5
New KG = (1,625,258 + 26,000) ÷ 26,435 = 62.46 ft

Need to interpolate to find out the MH1⁰ @ 73.5 ft draft and VCG 62.46

VCG 60	VCG 65	difference
3,285	978	= 2,307

MH1° @ 73.5 ft draft and VCG of 62.46 = 3,285 – [2,307 ÷ (65 – 60) × 2.46] = 2,149

Heel = moments ÷ MH1° = (200 × 47) ÷ 2,149
= 4.37 degrees to port

3. @ 65 ft draft KG = 65 ft
List = 2.5°
MH1° @ 65 ft draft and KG of 65 = 989
So for list 2.5° total moments = 989 × 2.5
= 2,472
Moments = weight × distance
Distance to move for zero list = moments ÷ weight
= 2,472 ÷ 25
= 98.9 ft

KM @ 65 ft draft = 67.24
So GM = KM – KG
= 67.24 – 65
= 2.24 ft *(GM is positive)*

Since GM is positive, seems okay to move the load.

4. At 70.5 ft draft with zero heel and trim. KG = 64.16 ft and displacement = 26,035

New KG and Displacement

Wt	×	VCG	=	V- moments
26,035	×	64.16	=	1,670,405
335	×	130	=	43,550

26,370 (new displacement) and draft = 73 ft
New KG = (1,670,405 + 43,550) ÷ 26,370 = 65 ft

At 73 ft draft and KG of 65 ft
MH1° = 971 ft-st MT1° = 1,970 ft-st

Heel = moments ÷ MH1°
= (335 × 27.5) ÷ 971
= 9.48 degrees to starboard

Heel in ft:
Tan of angle = heel in ft ÷ width of vessel
Tan of angle 9.48 = .166
Heel in ft = tan of angle × width of vessel
= .166 × 195
= 32.56 ft.
Heel on each side = 32.56 ÷ 2
= 16.28 ft (16.3)

Drafts

	Port Fwd	Port Aft	Stb Fwd	Std Aft
New draft	73	73	73	73
Heel	- 16.3	-16.3	+16.3	+16.3
	56.7	**56.7**	**89.3**	**89.3**

5. At 76 ft draft and KG of 65 ft
MH1° = 1,009 ft-st MT1° = 2,032 ft-st
Heel = moments ÷ MH1°
= (80 × 30) ÷ 1,009
= 2.37 degrees to starboard

Heel in ft:
Tan of angle = heel in ft ÷ width of vessel
Tan of angle 2.37 = .0415
Heel in ft = tan of angle × width of vessel
= .0415 × 195
= 8.10 ft.
Heel on each side = 8.10 ÷ 2
= 4.05 ft

Trim = moments ÷ MT1°
= (80 × 40) ÷ 2,032
= 1.57 degrees to aft

Trim in ft:
Tan of angle = trim in ft ÷ length of vessel
Tan of angle 1.57 = .027
Trim in ft = tan of angle × length of vessel
= .027 × 225
= 6.18 ft.
Trim on each side = 6.18 ÷ 2
= 3.1 ft

Drafts

	Port Fwd	Port Aft	Stb Fwd	Std Aft
New draft	76	76	76	76
Heel	- 4	-4	+4	+4
Trim	- 3.1	+ 3.1	- 3.1	+ 3.1
	68.9	**75.1**	**76.9**	**83.1**

6. a.

Wt	×	VCG	=	V- moments
26,300	×	60.06	=	1,579,578
267	×	54.08	=	14,439

26,567 (new displacement)
New KG = (1,579,578 + 14,439) ÷ 26,567 = 60 ft

b. **Sinkage = Wt ÷ TP1" = 267 ÷ 11.14 = 24" (2 ft)**

c. **New displacement = 26,567**
Corresponding mean draft = 74.5 ft

d. Heel = moments ÷ MH1°
= (267 × 95) ÷ 3,306
= **7.67 degrees to starboard**

e. *Change in heel in ft:*
Tan of angle = heel in ft ÷ width of vessel
Tan of angle 7.67 = .134
Heel in ft = tan of angle × width of vessel
= .134 × 195 = 26.26 ft.
Heel change on each side = 26.16 ÷ 2 = **13.13 ft**

f. **Trim = moments ÷ MT1° = (267 × 100) ÷ 4,317**
= 6.18 degrees to forward

g. *Change in trim in ft:*
Tan of angle = trim in ft ÷ length of vessel
Tan of angle 6.18 = .108
Trim in ft = tan of angle × length of vessel
= .108 × 225 = 24.36 ft.
Trim change on each side = 24.4 ÷ 2 = **12.2 ft**

h. *Drafts*

	Port Fwd	Port Aft	Stb Fwd	Std Aft
New draft	72.5	72.5	72.5	72.5
Sinkage	2	2	2	2
Heel	– 13.13	– 13.13	+ 13.13	+ 13.13
Trim	+ 12.2	– 12.2	+ 12.2	– 12.2
	73.57	**49.17**	**99.83**	**75.43**

7. **New KG and Displacement**

Wt	×	VCG	=	V- moments
25,965	×	61.2	=	1,589,058
135	×	132	=	17,820

26,100 (new displacement) and draft = 71 ft
New KG = (1,589,058 + 17,820) ÷ 26,100 = 61.56 ft

Need to interpolate to find out the MH1⁰ @ 71 ft draft and VCG 62.46

VCG 60	VCG 65	difference
3,239	961	= 2,278

MH1° @ 71 ft draft and VCG of 61.56 = 3,239 – [2,278 ÷ (65 – 60) × 1.56] = 2,528

Heel = moments ÷ MH1°
= (135 × 96) ÷ 2,528
= **5.12 degrees to starboard**

Heel in ft:
Tan of angle = heel in ft ÷ width of vessel
Tan of angle 5.12 = .089
Heel in ft = tan of angle × width of vessel
= .089 × 195
= 17.5 ft.
Heel on each side = 17.5 ÷ 2
= **8.75 ft**

Need to interpolate to find out the MH1⁰ @ 71 ft draft and VCG 61.56 ft

VCG 60	VCG 65	difference
4,213	1,936	= 2,277

MH1° @ 71 ft draft and VCG of 61.56 = 1,213 – [2,277 ÷ (65 – 60) × 1.56] = 3,502

Trim = moments ÷ MT1°
= (135 × 75) ÷ 3,502
= **2.89 degrees to aft**

Trim in ft:
Tan of angle = trim in ft ÷ length of vessel
Tan of angle 2.89 = .050
Trim in ft = tan of angle × length of vessel
= .050 × 225
= 11.25 ft.
Trim on each side = 11.25 ÷ 2
= **5.6 ft**

Drafts

	Port Fwd	Port Aft	Stb Fwd	Std Aft
New draft	71	71	71	71
Heel	- 8.75	- 8.75	+ 8.75	+ 8.75
Trim	- 5.6	+ 5.6	- 5.6	+ 5.6
	56.65	**67.85**	**74.15**	**85.35**

CHAPTER 4

Exercise, page 4.13

1. a. Max mooring tension for survival condition is 50% of breaking strength.
= 800,000 x .5
= 400,000 lb

b. Proof load is 66% of breaking strength.
= 800,000 x .66
= 528,000 lb

c. Max operation tension is 50% of proof load.
= 528,000 x .5
= 264,000 lb

2. a. $F_w = C_w \times V_w^2$
$F_w = .065 \times 30 \times 30$
$= 58.5$ Kips
$= 58.5 \div 2 = 29.25$ st

b. $F_c = C_c \times V_c^2$
$V_c^2 = F_c \div C_c$
$= 100 \div 37$
$= 2.7$
$= 1.64$ kt

c. 45 kips from chart
$45 \div 2 = 22.5$ st

3. Holding power of mooring system = Anchor holding power + Holding power of the chain or wire

Anchor holding power = Weight of anchor x the efficiency of the anchor
= 30,000 x 12 = 36,000

Holding power required = 400,000

Therefore, holding power of chain or wire = 400,000 – 36,000
= 40,000

Holding power of chain or wire Pc = $f \times L_c \times W_c$
$40{,}000 = .9 \times L_c \times 77.36$
Therefore, $L_c = 40{,}000 \div .9 \div 77.36$
= 575 ft

4.
$$S = \sqrt{\frac{2F \times Y}{W} - Y^2}$$

$$S = \sqrt{\frac{2 \times 300{,}000 \times 650}{77.36} - 650 \times 650}$$

S = 2,149 ft

where

F = mooring tension at the fairleader in lb
V = vertical component of the tension at the fairleader in lb
Y = vertical distance fairleader to seabed or height of the cable in ft
W = submerged weight of anchor cable in lb/ft in.
S = suspended length of anchor cable in ft

$F = H + (W \times Y)$
$H = F - (W \times Y)$
$H = 300{,}000 - (77.36 \times 650)$
= 249,716

where

F = mooring tension at the fairleader in lb
H = horizontal component of the tension at the fairleader in lb
Y = vertical distance fairleader to seabed or height of the cable in ft
W = submerged weight of anchor cable in lb/ft in.

$$X = \frac{H}{W}\, l_n \left[\frac{\left(\frac{F}{W} - S\right)}{\frac{H}{W}}\right]$$

$$X = \frac{249{,}716}{77.36}\, l_n \left[\frac{\frac{300{,}000}{77.36} - 2{,}149}{\frac{249{,}716}{77.36}}\right]$$

X = 2,015 ft

where

F = mooring tension at the fairleader in lb
V = vertical component of the tension at the fairleader in lb
H = horizontal component of the tension at the fairleader in lb
W = submerged weight of anchor cable in lb/ft in.
S = suspended length of anchor cable in ft
X = horizontal distance fairleader to touch down point of the anchor cable
l_n = natural log

5.
$$S = \sqrt{\frac{2F \times Y}{W} - Y^2}$$
$Y = 600 - 80 + 48$
$Y = 568$ ft
$$S = \sqrt{\frac{2 \times 300{,}000 \times 568}{77.36} - 568^2}$$

S = 2,021 ft

where

F = mooring tension at the fairleader in lb
Y = vertical distance fairleader to seabed or height of the cable in ft
W = submerged weight of anchor cable in lb/ft in.
S = suspended length of anchor cable in ft

6. Chain buoyant wt = 2,000 x 77.36 = 154,720 lbs
Anchor buoyant wt = 30,000 x .869 = 26,070
Total buoyant wt = 154,720 + 26,070 = 180,790 lbs

7. The optimum tension is the tension applied to all the anchors on completion of the test tension. When the anchors have optimum tension, they will at maximum operating offset reach, maximum operating tension.

8. Volume = $\pi r^2 h$

Volume of chain deployed = (2,000 x 90 x 10.2) ÷ 2000
= 918 ft^3

Height of stored chain deployed was = Volume ÷ π ÷ r^2
= 918 ÷ 3.14 ÷ (2.5 x 2.5)
= 46.7 ft

Vertical moments lost due to chain deployment = w x d
= (2,000 ft x 90 lb/ft) x {85 – (46.7÷2)}
= 11,097,000 ft-lbs
= 11,097,000 ÷ 2,000
= 5,548 ft-st

Vertical components gained

$$V = W \times S \text{ or } V = \sqrt{2F \times Y \times W - (Y \times W)^2}$$

$$S = \sqrt{\frac{2F \times Y}{W} - Y^2}$$

$$S = \sqrt{\frac{2 \times 140{,}000 \times (450 - 83.5 + 48)}{78.24} - 78.24^2}$$

$$S = \sqrt{\frac{280{,}000 \times (414.5)}{78.24} - 6{,}121}$$

$$S = \sqrt{1{,}483{,}384 - 6{,}121}$$

S = 1,215

V = W x S

V = 78.24 x 1,215

V = 95,061 lbs

where

F = mooring tension at the fairleader in lb

V = vertical component of the tension at the fairleader in lb

Y = vertical distance fairleader to seabed or height of the cable in ft

W = submerged weight of the anchor cable in lb/ft in.

S = suspended length of anchor cable in ft

ST = 95,061÷2,000 = 47.53 st

Vertical moments gained = w x d
= 47.53 x 48
= 2,281 ft- st

Correction for moments = Vertical moments lost – vertical moments gained
= 5,548 – 2,281
= 3,267 (Correction is minus, since vertical moments lost is greater than vertical moments gained.)

CHAPTER 5

Exercise, page 5.22

1. a. Down-flooding angle is between 35 to 45 degrees
   b. Angle of steady heel cause by the wind is 5 degrees

Angle	Righting arm GZ for angle from cross	Sin θ		KG		Correction for actual KG	Righting arm GZ—correction for actual KG		Displacement	Moments
5	8.5	0.087	x	65	=	**5.66**	2.85	x	20605	**58621**
10	18	0.173	x	65	=	**11.25**	6.76	x	20605	**139187**
15	29	0.258	x	65	=	**16.77**	12.23	x	20605	**251999**
20	40.5	0.342	x	65	=	**22.23**	18.27	x	20605	**376453**
25	53	0.422	x	65	=	**27.43**	25.57	x	20605	**526870**
30	65	0.500	x	65	=	**32.50**	32.50	x	20605	**669663**
35	72	0.573	x	65	=	**37.25**	34.76	x	20605	**716127**
40	72	0.642	x	65	=	**41.73**	30.27	x	20605	**623713**
45	70.5	0.707	x	65	=	**45.96**	24.55	x	20605	**505750**

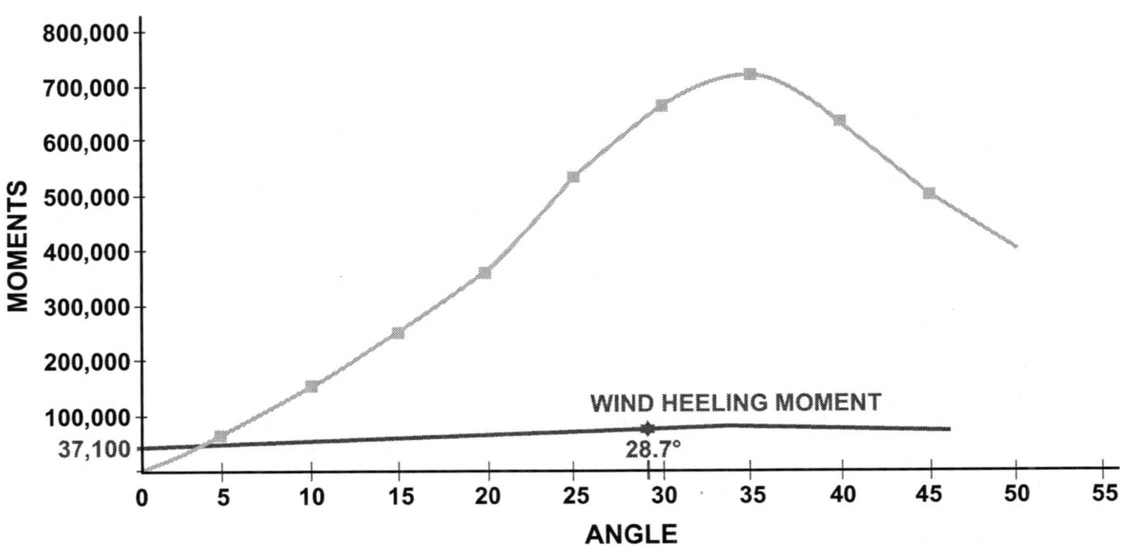

2. a. Down-flooding angle is between 30 to 35 degrees
   b. Angle of steady heel cause by the wind is 5 degrees

Angle	Righting arm GZ for angle from cross	Sin θ		KG		Correction for actual KG	Righting arm GZ—correction for actual KG		Displacement	Moments
5	6	0.087	x	55	=	**4.79**	1.22	x	25938	**31515**
10	12	0.173	x	55	=	**9.52**	2.49	x	25938	**64456**
15	18	0.258	x	55	=	**14.19**	3.81	x	25938	**98824**
20	23	0.342	x	55	=	**18.81**	4.19	x	25938	**108680**
25	33	0.422	x	55	=	**23.21**	9.79	x	25938	**253933**
30	40.5	0.500	x	55	=	**27.50**	13.00	x	25938	**337194**
35	43.5	0.573	x	55	=	**31.52**	11.99	x	25938	**310867**
40	43.5	0.642	x	55	=	**35.31**	8.19	x	25938	**212432**
45	43.5	0.707	x	55	=	**38.89**	4.62	x	25938	**119704**

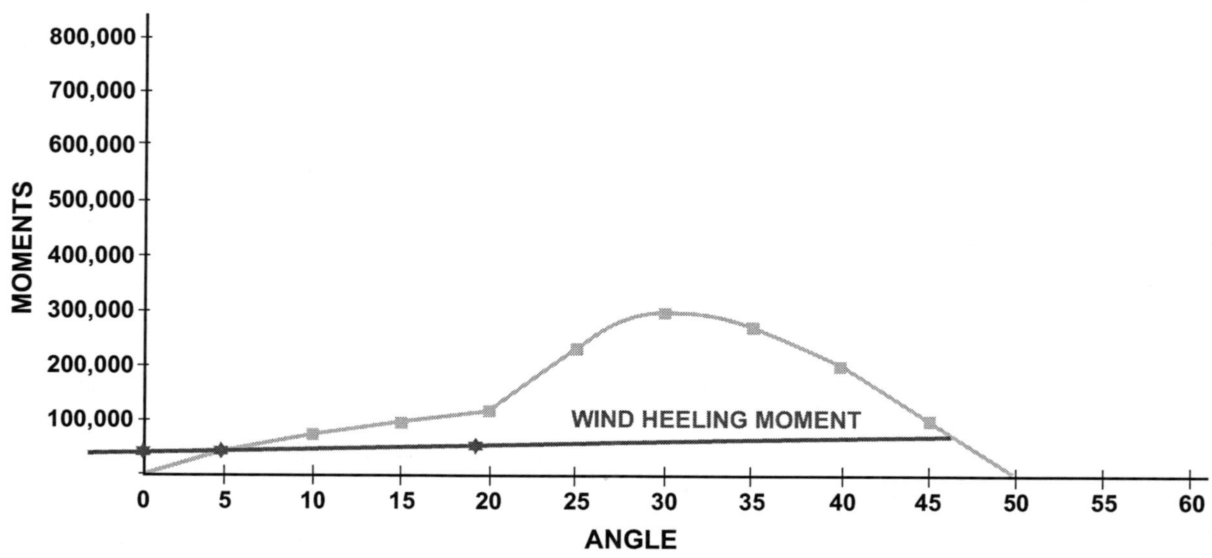

3. Ingress of seawater = 649 st of seawater (from table 3.6, chapter 3)

Moment = weight x distance
= 649 x 110
= 71,390 ft-st

Heeling moment due to flooding

Angle	Sin θ		Moment		Heeling Moment
5	0.996	x	71390	=	**71104.44**
10	0.984	x	71390	=	**70247.76**
15	0.965	x	71390	=	**68891.35**
20	0.939	x	71390	=	**67035.21**
25	0.906	x	71390	=	**64679.34**
30	0.866	x	71390	=	**61823.74**
35	0.819	x	71390	=	**58468.41**
40	0.766	x	71390	=	**54684.74**
45	0.707	x	71390	=	**50472.73**

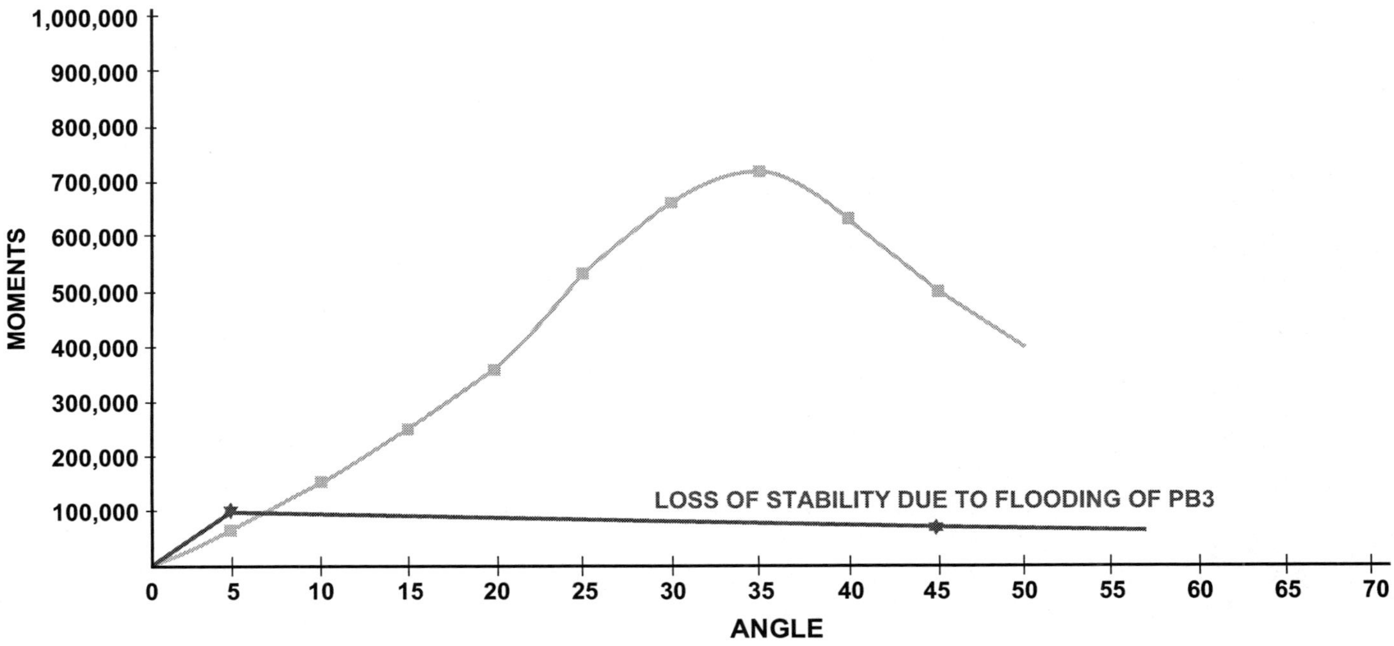

4.

Angle	Heeling moment from question 1		Heeling moment due to flooding (question 3)		Heeling Moment
5	**58621**	-	71104	=	**-12483**
10	**139187**	-	70247	=	**68940**
15	**251999**	-	68891	=	**183108**
20	**376453**	-	67035	=	**309418**
25	**526870**	-	64679	=	**462191**
30	**669663**	-	618223	=	**51440**
35	**716127**	-	58468	=	**657659**
40	**623713**	-	54684	=	**569029**
45	**505750**	-	50472	=	**455278**

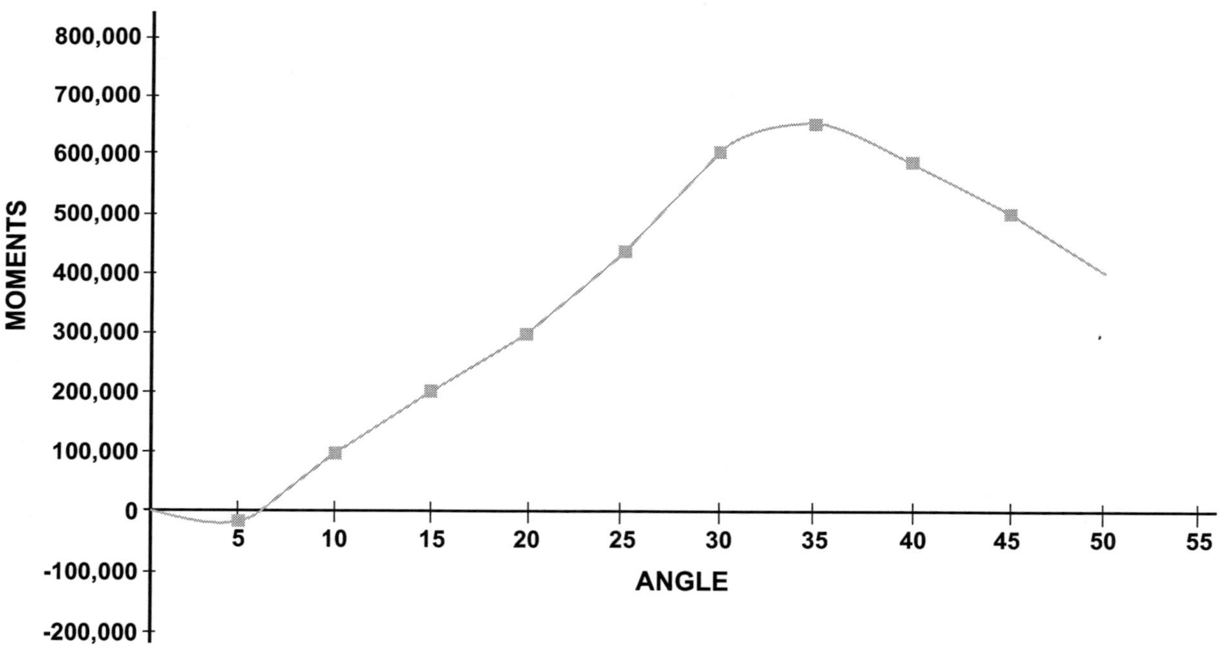

5. If a correction is made for the KG value after the flooding of PB3 (with a VCG of 10.5 ft), the KG of the rig will be less then the original KG. Hence the GM increases giving the rig more stability. Therefore the angle of list will be less.

6. True. KG will change, hence the GM will change. If too much ice loading occurs, this could result in the KG reaching the Maximum KG value or going past it. This will result in the GM value to be too low.

7. Rigs are prohibited to operate in *shaded* areas, as there is a possibility of multiple (2 or more) compartments flooding, after a collision or a similar incident. This could result in negative stability.

8. The stability curve will have a longer range, due to the increase in the freeboard.

   Note:
   BM = I/V = (Length x Breadth^3) ÷ (12 x volume)
   The righting moment curve values will also be greater.

9. Tangent angle = opposite ÷ adjacent
   = 16 ÷ 38.5
   = .4155
   Corresponding angle = 22.56 degrees

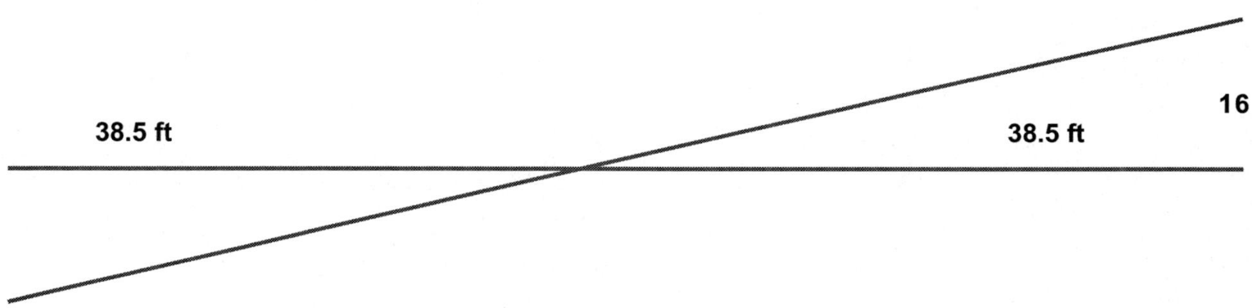

CHAPTER 6
Exercise, page 6.11

1. Use the attached hydrostatic properties and tables.
   The Rig Type 2A in deep tow condition accidentally floods a previously empty PDW2 to half depth with fresh water.
   VCG of the tank is 20 ft aft of LCF and 108 ft to port of centerline.
   Towing draft = 55 ft
   Initial KG = 62 ft
   LCF is at the centerline of the rig
   Length between draft marks is 225 ft.
   Width between the draft marks is 195 ft.
   Calculate the final draft of each column.
   Calculate final GM_{wet}, and KG longitudinal and KG transverse.
   (see next page)

Initial draft = 55 ft **Displacement = 23,965 st** **KG = 62 ft**

Wt increase due to flooding tank half full = $\frac{550}{2}$ = **275 st** VCG half tank = $\frac{10.5}{2}$ = **5.25 ft**

Moments

	Wt	VCG	V-Moms	TCG	T-Moms	LCG	L-Moms
Displ.	23,965	62	1,485,830	0	0	0	0
Wt added	275	5.25	1,443.75	-108	-29,700	-20	-5,500
Total	**24,240**	**61.36**	**1,487,273.75**		**-29,700**		**-5,500**

New Displacement = **24,240** New Draft = **57 ft** Increase = **57 – 55** = **2 ft all around**

FSC = $\frac{\text{FSM}}{\text{Displ.}}$ FSM for tank = 2,968 FSC = $\frac{2{,}968}{24{,}240}$ = **.122 ft**

KG_{wet} = KG + FSC KG_{wet} = 61.36 + .122 = **61.5 ft**

Heel

MH1 degree @ VCG 61.5 ft

VCG 60	VCG 70	Difference	MH1 degree @ VCG 61.5 ft
3,269	1,154	2,115	$3{,}269 - \frac{(2{,}115 \times 1.5)}{10}$ = **2,951**

Heel Degrees = $\frac{29{,}700}{2{,}951}$ = **10.06 degrees**

Heel in ft = Tan angle = Heel / Breadth Heel = Tan angle x Breadth

Tan 10.06 degrees = **.177** Heel = .177 x 195 = **34.5 ft** Each side 34.5 / 2 = **17.25 ft**

Trim

MT1 degree @ VCG 61.5 ft

VCG 60	VCG 70	Difference	MT1 degree @ VCG 61.5 ft
4,081	1,966	2,115	$4{,}081 - \frac{(2{,}115 \times 1.5)}{10}$ = **3,763**

Trim Degrees = $\frac{5{,}500}{3{,}763}$ = **1.46 degrees**

Trim in ft = Tan angle = Trim / Length Trim = Tan angle x Length

Tan 1.46 degrees = **.025** Trim = .025 x 225 = **5.6 ft** Each side 5.6 / 2 = **2.8 ft**

Drafts

	PF	PA	SF	SA
Initial	55	55	55	55
Increase due flooding	2	2	2	2
Heel	17.25	17.25	-17.25	-17.25
Trim	-2.8	2.8	-2.8	2.8
	71.45	**77.05**	**36.95**	**42.55**

Free surface corrections

FSC T = $\frac{2{,}968}{24{,}240}$ = **.122** **FSC L** = $\frac{1{,}767}{24{,}240}$ = **.072**

@ Draft 57 ft

KMT =	67.73	**KML** =	69.65
KG =	-61.35	**KG** =	-61.35
FSC T =	-0.122	**FSC L** =	-0.072
GM T =	**6.258**	**GM L** =	**8.228**

2. Rig Type 2A is WOW with a deck load of 2,200 st at a draft of 69.9 ft. VCG = 56.2 ft. Damage occurs to port aft 30 ft diameter column between 35 ft and 65 ft flats. The permeability of the flooded compartment is 95%. Calculate the final draft for each column and the new KG with a compartment permeability of 95%.

Displacement at 69.95 ft = 25,958 st.
Length between draft marks = 225 ft. Width = 195 ft.
Center of the flooded column is 97.5 ft from LCF.
Center of the flooded column is 112.5 ft from TCF.
COF is at the centerlines.

Initial draft = 69.95 ft **Displacement = 25,958 st** **KG = 56.2 ft**

The 30 ft column is filled with seawater from 35 ft to 65 ft. Permeability is 95%.

So number of cubic feet that can be filled with water = pi x radius x radius x height x permeability
= 3.1415927 x 15 x 15 x 30 x .95 = **20,145.463 cu ft**
This corresponds to short tons = 20,145.463 / 31.22 = **645.68 st**

Moments

	Wt	VCG	V-Moms	TCG	T-Moms	LCG	L-Moms
Displ.	25,958	**56.2**	1,458,839.6	0	0	0	0
Wt added	645	50	32,250	97.5	62,887.5	112.5	72,562.5
Total	**26,603**	**56.05**	**1,491,089.6**		**62,887.5**		**72,562.5**

New Displacement = 26,603 KG = 56.05 ft After interpolation draft = 74.75 ft

Heel

MH1 degree @ VCG 56.08 ft	VCG 50	VCG 60	Difference	MH1 degree @ VCG 56 ft
	7,955	3,312.5	4,642.5	$\frac{7{,}955 - (4{,}642 \times 6)}{10}$ = **5,169**

Heel Degrees = $\frac{62{,}887.5}{5{,}169}$ **= 12.16 degrees**

Heel in ft = Tan angle = Heel / Breadth; Tan 12.16 degrees = **.215**

Heel = Tan angle x Breadth; Heel = .215 x 195 = **42 ft**

Each side 42 / 2 = **21 ft**

Trim

MT1 degree @ VCG 56.08 ft	VCG 50	VCG 60	Difference	MT1 degree @ VCG 56 ft
	8,967	4,324	4,643	$\frac{8{,}967 - (4{,}643 \times 6)}{10}$ = **6,181**

Trim Degrees = $\frac{72{,}562.5}{6{,}181}$ **= 11.7 degrees**

Trim in ft = Tan angle = Trim / Length; Tan 11.7 degrees = **.207**

Trim = Tan angle x Length; Trim = .207 x 225 = **46.6 ft**

Each side 46.6 / 2 = **23.3**

Drafts

	PF	PA	SF	SA
Draft	74.75	74.75	74.75	74.75
Heel	21	21	-21	-21
Trim	-23.3	23.3	-23.3	23.3
	72.45	**119.05**	**30.45**	**77.05**

3. Rig Type 2A is loaded as shown on the attached load form summary page. Calculate the final drafts for each column if the port aft column is damaged and floods between the 51 ft and 65 ft horizontal flats.

Initial draft = 83.5 ft **Displacement = 27,773 st** **KG = 59.23 ft**

Moments

	Wt	VCG	V-Moms	TCG	T-Moms	LCG	L-Moms
Displ.	27,773	59.23	1,644,994.79	0	0	0	0
Wt added	260		15,118		25,370		29273
Total	**28,033**	**59.22**	**1,660,112.79**		**25,370**		**29273**

New Displacement = 28,033 **New Draft = 85.5** **KG = 56.08 ft**

Heel

MH1 degree @ VCG 60 ft

VCG 50	VCG 60	Difference
8,608	3,714	4,894

MH1 degree @ VCG 56 ft $\frac{8{,}608 - (4{,}894 \times 9.2)}{10}$ = **4,105**

Heel Degrees = $\frac{25{,}370}{4{,}105}$ = **6.18 degrees**

Heel in ft =

Tan angle = Heel / Breadth

Tan 6.18 degrees = **.108**

Heel = Tan angle x Breadth

Heel = .108 x 195 = **21 ft**

Each side 21 / 2 = **10.5 ft**

Trim

MT1 degree @ VCG 56.08 ft

VCG 50	VCG 60	Difference
9,724	4,830	4,894

MT1 degree @ VCG 56 ft $\frac{9{,}724 - (4{,}894 \times 9.2)}{10}$ = **5,221**

Trim Degrees = $\frac{29{,}273}{5{,}221}$ = **5.60 degrees**

Trim in ft =

Tan angle = Trim / Length

Tan 5.6 degrees = **.098**

Trim = Tan angle x Length

Trim = .098 x 225 = **22.06 ft**

Each side 22.06 / 2 = **11**

Drafts

	PF	PA	SF	SA
Draft	85.5	85.5	85.5	85.5
Heel	10.5	10.5	-10.5	-10.5
Trim	-11	11	-11	11
	85	**107**	**64**	**86**

CHAPTER 7

Exercise, page 7.10

1. a. To allow the rig to float at a specific draft.
   b. To be able to trim the rig—maintain 0 heel and trim.
   c. Test soil strength in jackup operations prior to drilling.
   d. Counteract accidental flooding.

2. Main components of ballast system.
   a. ballast tanks
   b. ballast piping system
   c. pumps
   d. valves
   e. sea chests
   f. overboard discharge
   g. vent lines
   h. sounding systems

3. It is necessary to strip almost empty tanks to prevent too much free surface which causes a virtual loss of stability.

4. a. Close ballast pump discharge valve.
   b. Shut down pump immediately after discharge valve is closed.
   c. Close sea chest valve.
   d. Close tank manifold valves.

5. Block valve is installed to
   a. isolate forward and aft section of valves to prevent equalization between forward and aft tanks in case of leaking valves.
   b. to have redundacy in case of pump failure.
   c. to enable to transfer ballast from forward to aft and vice versa.

6. By causing it to slam across its seat by improper sequence of closing the valves after ballasting/deballasting operation. If check valve breaks/shatters, the parts could end up in the pump house causing serious damage/destruction of the pump.

7. To have a dual possibility to close the intake at the sea chest and have a minimum length of "pipe" exposed to seawater pressure in ballasted condition.

8. True. However, in a damage stability condition with emergency power, only then only one pump can be used at the time.

9. The correct shutdown sequence procedure for deballasting is even more important than for ballasting, because of the increased water pressure caused by the height of the water column. Secondly, valve slamming is more severe due to opposite flows.

10. No boats alongside during ballasting/deballasting as a possible collision could cause a two compartmental flooding from which the rig could not survive.

11. a. Operation possible separately or in conjunction with main ballast system.
    b. Independent operation with main power (only one pump with emergency power).
    c. Can be used to counteract flooding by deballasting if main system is rendered useless.
    d. Can deballast to survival draft.
    e. In case of extreme trim situations causing suction lift problems for main ballast system.

12. a. Increase in WPO? which affects the stability.
    b. If free flooding then we do have a large free surface moment over a great large transverse distance.

13. Fail safe is important that the sea chest valve closes automatically in case of power failure.

14. Presence of air bubbles in the fluid—which causes pitting on the pump impeller.

CHAPTER 8

Exercise, page 8.10

1. • Loads that tend to force the structure apart
   • Also (transverse and diagonal) loads to compute stress ranges per foot/meter wave height so that fatigue life of critical members can be checked.

2. • Brace connections at column shells
   • Vent holes
   • Drains and hydrophone wells
   • Cable way penetrations
   • Butt welds at K joints
   • Tube to tube connections

3. A jackup preloads to test soil condition to withstand the extra force on the leeward leg in case of storm conditions. This is done by filling the so-called preload tanks and holding this weight for a duration of at least six hours.

4. • Statical loads such as weight of rigs and component buoyancy force and water pressure.
   • Dynamic loads caused by environmental forces such as roll, pitch, yaw, sway, heave, and surge.

5. Fatigue is the tendency to fracture under cyclic stress. High loads and increased cycles in the fatigue mode may reduce the effective strength of the material.

6. Stress—produced with an axial pulling within the material; expressed in psi or kg/cm^2.
   Strain—elongation in specimen represents strain; expressed in in./in. or cm/cm.

7. Sagging—occurs when wave crests are at extreme ends—giving ends more buoyancy than midships; bending force amidship.
   Hogging—occurs when wave crest is amidships causing bending force at ends.

8. • Legs–spud can connection
   • Jackhouse connection to the hull
   • Hold-down claw for the cantilever

9. The load curve of a drilling vessel represents shear and bending forces for sagging and hogging—difference between weight and buoyancy ordinates.

10. The utilization factor of .75 represents that the material is exposed to 75% of allowable maximum stress. Stressing the material beyond 100% will cause overstress and may result in failure.

11. The yield point is the point of proportional elastic limit where material will elongate without almost any additional load.

12. The point where the steel is permanently deformed is called the proportional elastic limit.

13. Half gravity loads are loads to calculate the approximate forces caused by pitch and roll. These loads control design of lower hulls, columns, vertical bracings, and main deck girders.

14. Asymmetric loads are **not** allowed on the rig except in emergency damage stability conditions which is unintentional flooding of one or more compartments.

15. Twisting force is caused by waves with simultaneous crests at two diagonal corner columns and a wave trough at the two opposing corner columns.

16. The endurance limit is the maximum stress that can be applied to a structure for an indefinite service of life.